Kirsten Winter

Model Checking Abstract State Machines

Kirsten Winter

Model Checking Abstract State Machines

VDM Verlag Dr. Müller

Imprint

Bibliographic information by the German National Library: The German National Library lists this publication at the German National Bibliography; detailed bibliographic information is available on the Internet at http://dnb.d-nb.de.

Any brand names and product names mentioned in this book are subject to trademark, brand or patent protection and are trademarks or registered trademarks of their respective holders. The use of brand names, product names, common names, trade names, product descriptions etc. even without a particular marking in this works is in no way to be construed to mean that such names may be regarded as unrestricted in respect of trademark and brand protection legislation and could thus be used by anyone.

Cover image: www.purestockx.com

Publisher:
VDM Verlag Dr. Müller Aktiengesellschaft & Co. KG , Dudweiler Landstr. 125 a, 66123 Saarbrücken, Germany,
Phone +49 681 9100-698, Fax +49 681 9100-988,
Email: info@vdm-verlag.de

Zugl.: Berlin, TU, Diss., 2001

Produced in USA and UK by:
Lightning Source Inc., La Vergne, Tennessee, USA
Lightning Source UK Ltd., Milton Keynes, UK
BookSurge LLC, 5341 Dorchester Road, Suite 16, North Charleston, SC 29418, USA

ISBN: 978-3-639-01891-2

Acknowledgements

First of all, I would like to thank my supervisor Stefan Jähnichen who employed me as a PhD student at GMD FIRST. His open and personal approach towards his group strongly influenced my development from a student to a researcher. I was given much scope in choosing my own direction for my PhD topic. Moreover, the opportunity of working on other research projects encouraged me to consider many areas of related work. My experience at GMD FIRST has provided me with a broad background in the field of formal methods that will be useful in the future, I am sure.

Next, I would like to thank Wolfgang Reisig for spontaneously agreeing to be my second supervisor. His discussions, often very critical, on drafts of this work helped improve it alot. I very much appreciated the time he took with me to go through his comments. Something one must learn when doing a PhD is to defend their results. I guess, a big part of this I learned on the other side of the road, at the Humboldt University.

Also, Egon Börger's support has had a great impact on this thesis. He encouraged me to write my first publication which is, I believe, the most difficult one and gave me many helpful corrections and much feedback on my first attempt at this paper. While developing my ideas on model checking ASM, he encouraged me alot by showing continuous interest in my results.

What is the use of a fully automatic model checker if it needs a manual transformation of the models? Giuseppe Del Castillo encouraged me to think about implementation. I was not always happy when fighting with the intricacies of ML but in the end I am very grateful for this cooperation.

Sofiene Tahar earns my gratitude for his useful comments on my use of MDGs. I hope to continue this cooperation in the future.

The research group of GMD and the Technical University Berlin in which I "grew up" provided a nice atmosphere for discussions. I thank Jochen Burghardt, Carsten Sühl, Florian Kammüller, Steffen Helke, Thomas Santen, Matthias Anlauff and Stephan Herrmann for their helpful comments and their personal support. Especially, I would like to thank Stephan for his companionship through the first years of my PhD.

Many thanks to Graeme Smith for fighting with the drafts and, as a consequence, for giving me "lessons" in English writing. His sometimes differing point of view led to many useful discussions on formal notations. Apart from this professional support, I want to thank him for keeping me cheery through these last winter months and for the bowls of pasta that ensured survival in the late working hours.

Finally, I would also like to thank my mother and my siblings for the love and the sunshine they bring into my life.

Zusammenfassung

Abstract State Machines (ASM) ist eine formale Spezifikationssprache, die es erlaubt, auf einem hohen Abstraktionsniveau zu modellieren. Sie ist gut geeignet für verschiedenartigste Anwendungen. Computerbasierte Werkzeugunterstützung ist in Form von Editoren, Typecheckern und Simulatoren vorhanden. ASM ist außerdem in die Logiken zweier Theorembeweiser eingebettet worden, die interaktives Beweisen unterstützen.

Diese Dissertation hat zum Ziel, die vorhandene Werkzeugunterstützung um einen vollständig automatisierten Ansatz zu erweitern, das sogenannte *Modelchecking*. Modelchecking ist die vollständige Suche im Zustandsraum des zu untersuchenden Systemmodells. Die Algorithmen arbeiten automatisch und bedürfen keiner Interaktion mit dem Benutzer oder der Benutzerin. Als zentrale Werkzeugumgebung wird die *ASM Workbench* verwendet, in die ein Modelchecker integriert wird. Diese Werkzeugumgebung wird mit einer allgemeinen Schnittstelle in Form einer Zwischensprache versehen. Diese allgemeine Schnittstelle dient als Grundlage zur Anbindung verschiedener Modelchecker, die auf einfachen Transitionssystemen basieren. Mit der Transformation der abstrakten Spezifikationssprache ASM in die sehr einfachen Sprachen von vollautomatischen Werkzeugen schlägt diese Arbeit eine Brücke über den tiefen Graben zwischen Modellierungssprachen, die auf hohem Niveau beschreiben, und Werkzeugsprachen, die algorithmisch einfach zu behandeln sind.

Auf der Basis der allgemeinen Schnittstelle werden im Rahmen dieser Dissertation zwei konkrete Schnittstellen entwickelt: von der ASM Workbench zum Modelchecker SMV und zum MDG-Paket. Ersteres Werkzeug wird häufig eingesetzt und implementiert symbolisches Modelchecken für CTL-Formeln.

Die Anbindung zum Modelchecker SMV ist implementiert und zwei Fallstudien wurden mittels der entwickelten Transformation und dem SMV Werkzeug untersucht. Die Ergebnisse zeigen die Anwendbarkeit aber auch die Grenzen dieses Ansatzes auf. Diese Grenzen motivieren die Entwicklung der zweiten Schnittstelle zum MDG-Paket. Dieses Paket umfaßt die Basisfunktionalität für symbolisches Modelchecking auf der Grundlage von Multiway Decision Graphs (MDGs). Diese Graphenstruktur erlaubt die Repräsentation von möglicherweise unendlichen Transitionssystemen mit Hilfe von abstrakten Sorten und Funktionen. Sie bietet daher ein einfaches Mittel, um die Abstraktion von Modellen zu unterstützen. Außerdem erlaubt dieser Ansatz, Annahmen über das zu untersuchende System und seine Umgebung in temporaler Logik zu spezifizieren. Die resultierende logische Modellierung wird anschließend mit dem Systemmodell vereinigt und führt zu einer vollständigeren Modellierung. Damit erweitert diese zweite Schnittstelle die Grenzen der vollautomatischen Analyse, die Modelchecking zur Verfügung stellt.

Contents

Chapter 1

Introduction

Why use Formal Methods?

Information technology has had a growing market for many years. Since the situation in this market is very competitive, the development of computer systems is forced to maximise productivity. In many cases, this is paid for with a decrease in quality. Quality, however, is an essential economic factor whenever liability has to be assumed by the producer. This is the case for hardware production where a faulty chip that is launched can cause a huge financial loss for the producer, as well as for safety-critical systems whose reliability has to be certified before use. Software engineering methods, in general, aim to improve the quality of software and system development. In particular, formal methods focus on reliability and correctness of systems by using a mathematical framework.

The benefits of formal methods have been discussed many times in the literature. Production costs can be decreased greatly if errors can be detected in an early phase of design. Accidents, involving major damage and even loss of lives, can be avoided if systems are working correctly. Nevertheless, the use of formal methods is limited in practice since the tradeoff between productivity and reliability is still too large. This work aims at improving the applicability of the formal approach.

Reliability of a system can be improved by modelling a design of the system before implementing the code, validating its appropriateness and verifying its correctness. Validation is done with respect to given informal requirements of the system. The designer or the user inspects if the model properly reflects the expected behaviour of the system. Verification is done with respect to formal requirements of the system. A formal requirement is a model of the system in a formal notation described at some higher level. Being in a formal framework, verification can be done using mathematical proofs.

Formal methods deal with modelling, validation and verification based on a formal notation. The research in this area has developed on two different levels: the language and the tool level, each of them with a particular focus.

The language level provides support for the user when modelling a system. That is, formal languages aim to bridge the gap between the informal and formal description. They have a precisely defined semantics in order to prevent ambiguities. They should be simple to learn and use, should lead to understandable models that allow inspection, and should be expressive enough for modelling various problems. To treat large systems, structuring mechanisms should support the design of well-structured systems. Additionally, abstraction or refinement techniques should be available to support modelling on different levels of abstraction. At a high level of abstraction, the model has reduced complexity because technical details of a possible realisation are left open. The notion of *high-level languages* is used to distinguish languages, capable of such high-level descriptions, from low-level languages such as programming languages. Many formal languages exist that fulfil the given criteria for suitable language support for modelling of large systems. One representative among them is the notation of Abstract State Machines (ASM) that is introduced in [Gur95, Gur97].

The tool level aims at tool support for the validation and verification tasks. Validation can be supported by simulators and animators. Verification is facilitated by theorem provers, proof checkers, model checkers, satisfiability checkers and so on. Tool support is recommendable when bringing formal methods into practice since many tasks can be facilitated by automatisation. Moreover, for the verification task, a tool-driven proof is usually more reliable. Manual proofs may be less tedious but they are often more error prone than mechanical proofs. Often, they are not seen to be fully reliable. Wolper (in [Wol97]) even draws the line between "weak" and "strong" formal methods according to the tool support provided for a given language.

At both levels of formal methods development many useful results are available: there are adequate high-level languages for modelling and manual analysis, and appropriate tools that facilitate the validation and verification tasks. In most cases, however, there is a gap between the kinds of support: The more facilities a language has, the more difficulties arising for the development of tools. The more automatisation that is available, the lower the level of the corresponding modelling language. There is a discrepancy between tools that are easy to use and languages that nicely support the modelling task. In fact, support on both ends of the formal methods spectrum is needed for them to be of practical relevance in the future. It is necessary to bridge this gap and relate both issues, language support and tool support. Basically, this is a transformation task that can be automated to a large extent due to the formal semantical foundation. This work aims to contribute with a solution for a particular formal language and a certain kind of tool support for verification. The title "Model Checking Abstract State Machines" summarises our goal.

Why use Abstract State Machines?

There are many formal specification languages that could be chosen for this work. We choose ASM to make preliminary experiences for the following reasons.

ASM is used as a modelling language in a variety of domains, e.g., embedded systems, protocols, hardware specifications, semantics of programming languages (a bibliography is given at [Hug]). It has been used both in academic and industry contexts. The wide group of users shows that there is interest in the language and, consequently, there is an interest in tool support. In fact, we can point out at least two research groups that are already using the results of this work (see [SMT01, GR01]).

A second reason is that ASM models transition systems in a simple and uniform fashion and give these transition systems an operational semantics. Many model checkers that are available are based on transition systems. A transformation from ASM into these transition-system-based model-checker languages can be done without losing properties of the original model. At a first glance, model checking appears to be a promising approach.

Why use Model Checking?

Verification tools can be classified into *interactive* and *automatic* tools. Theorem provers, at the current state of art, are not fully automatic for their usual tasks.

They are driven through user interaction. As a consequence, experts are needed that are familiar with logic and the particular proof system that underlies the prover. The proof task turns out to be very time and cost intensive if it can be completed at all.

Model checking tools, in contrast, do not provide a proof of correctness but rather execute an exhaustive search for errors in the state space of a model. It is an exhaustive test over all possibilities. This search can be done fully automatically. As a result, the user gets an answer that the checked requirement is either satisfied in the model or violated and, in this case, an example shows in which situation the violation may happen.

The process of searching a state space terminates only if this state space is finite. It will terminate in a reasonable span of time if the state space is finite and also small enough. Finiteness is a clear attribute. In contrast, the term "small enough" is very vague. It describes a limit that is not fixed but can be pushed by either optimising the model checking technique that is used or by "optimising" the model by reducing the size of the treated state space.

For small systems, model checking is a "press-button" task that provides a result almost for free: no expertise is needed, no user interaction is requested. This is very appealing for industry.

What else to use?

At a first glance, the tradeoff between reliability provided by model checkers and productivity appears acceptable. At a second glance, when actually using model checking, it becomes more obvious that this technique cannot fully substitute the technique of provers. Model checking is easy to apply only for systems with a small state space. Experiences show that model checking is not applicable for usual size models if complex data structures are involved. This drawback becomes particularly visible when analysing software systems in contrast to hardware systems.

The verification task for software is two-folded: analysing the control flow and analysing the data part. Often, the former task can be solved by a model checker but the latter needs the generality of theorem provers. Therefore, the overall goal for tool support is to exploit automatic techniques as far as possible and treat remaining proof tasks interactively if necessary. This framework of tool support needs a methodology for separating control from data issues. To apply model checking, a method is needed for filtering out the complex data parts of a model. One means of achieving this is *abstraction*.

It is important to understand that this notion of abstraction aims at a state space reduction. Modelling on a high level of abstraction, in contrast, aims to reduce the size of the model in terms of its appearance. Abstract models in the latter sense that appear simple to the user may actually involve an infinite state space.

In the context of model checking, abstraction allows the developer to skip information that is not relevant to a particular requirement that is to be checked. Abstracting from data that is irrelevant for the requirement results in a model that is smaller in terms of its state space and a checking process that is less time and space consuming. The problems are how to find out which information is not needed for checking a particular requirement, how to derive an abstract model once we have pointed out the irrelevant data, and how to proof that the abstract model preserves the properties of the original model. If we get model checking results for the abstract model what do they imply for the concrete model?

At this point, model checking turns out to be more complicated than a simple "press-button" facility. Much research is currently investigating how abstraction can be supported by tools in order to apply model checking to a wider range of problems. The work in this thesis aims at exploiting one suggested approach for deriving and checking abstract models.

Overview

According to the introduction so far, we want to provide ASM with a model checking tool and extend the limits of this approach. This work has two foci: a transformation of the high-level formal notation ASM into a lower-level language suitable for a model checker, i.e., a general interface, and its extension to provide a means for abstraction.

The thesis is divided into five parts:

i.) We describe the notation of ASM in Chapter 2. In order to give a first impression, a small example is given in the beginning that shows what ASM looks like. This is followed by the full definition of the syntax and semantics of the formalism. In Section 2.3, we briefly describe the ASM Workbench (see [Cas00]) which is the tool framework for ASM that we use as a core tool for our interface.

ii.) We present the field of model checking in Chapter 3. The introduction of this chapter starts with an overview of various techniques that are developed in the last decade. Section 3.3 and 3.4 focus on two particular approaches that we want to use for checking ASM: model checking of branching time temporal logic and linear time temporal logic. The theory of these two model checking approaches is recalled in detail here since its understanding allows us to motivate the design decisions for the current and future development of this work. In particular, Chapter 4 explains the data structure of decision diagrams which are the backbone of the model checking tools we use.

iii.) Chapter 5 describes the core of this work: applying model checking to ASM. In the first place, this chapter gives a general picture of how we think the approach is useful in practice. This is followed by a discussion about problems that arise in Section 5.1. Section 5.2 introduces the transformation algorithm from the ASM specification language into an intermediate language. This intermediate language provides a general interface to tools that are based on transition systems. We give a proof of correctness of the transformation on the basis of the ASM semantics given in Chapter 2.

iv.) The SMV tool ([McM93]) is a model checker that fits well into our framework, because it is a tool that operates on a transition system. Chapter 6 gives a short overview of the tool and its input language and presents the transformation from our intermediate language of ASM into the SMV input language. In Chapter 7 we describe, with two case studies, how we can model check ASM models by means of the SMV tool. Section 7.1 summarises our experiences with the FLASH cache coherence protocol (c.f., [Hei95, KOH$^+$94]). It provides a nice example of a successful debugging process. The second case study, the Production Cell (c.f., [LL95]), points out the limitations of the SMV approach. This is described in Section 7.2. It is shown why embedded systems, such as the Production Cell, cannot be handled properly by SMV.

v.) The experiences with the Production Cell motivate the development of an interface to *Multiway Decision Graphs* (MDGs). The foundation of MDGs and a basic functionality for MDG model checking has been developed at the University of Montreal and Concordia University, Canada (see [CCL$^+$97, CZS$^+$97]). MDGs provide a simple means for abstraction that can be nicely exploited for ASM. Chapter 8 motivates this second interface in detail. In particular, Section 8.3 describes the notion of abstraction and how this notion can be easily applied for abstracting ASM models. The central idea is to introduce abstract sorts into the framework of ASM which can be done on the syntactical level of the ASM language. This extension of the language makes it necessary to extend the generic transformation algorithm from the ASM language into the intermediate language. The extended transformation is defined in Section 8.4. As a second step, this section describes the mapping from the extended intermediate language into MDGs.

The outcome of this work is firstly an interface from the ASM Workbench to the model checker SMV. The implementation is optimised and validated through our use for the case studies that are described here. [SMT01] describes another application of this interface for hardware verification. This first interface was developed in cooperation with Giuseppe Del Castillo who developed the ASM Workbench. His insights into the internals of the Workbench were essential for implementing the interface.

Secondly, we provide an interface from the ASM Workbench to the MDG-Package. Through this second interface we extend the limits of model checking for ASM. We provide a simple means of generating an abstraction if the ASM model is too complex. Our implementation generates Prolog code for producing a representation of the ASM model which is suitable as input to a model checker based on MDGs. The realisation of an MDG model checker tailored for ASM is beyond the scope of this thesis but it is planned as future work in cooperation with Concordia University.

Chapter 2

Abstract State Machines

2.1 Introduction: Formalism and Methodology

Abstract State Machines (ASM) ([Gur95], [Gur97]) are a means of formalising algorithms or systems in a state-based way. With this notation we can formalise the possible *states* and *state transitions* of the system. That is, we can express a system structure as well as its dynamics within the same notational framework.

We use many-sorted first-order structures for describing the states of a system. These structures consist of the universes (or domains), comprising the entities our systems deals with, and declare functions over these universes (a simple example is given in Section 2.1.1). Universes and functions comprise the *state space*. The evaluation of functions depends on the current state if they are *dynamic*, i.e., a current state is determined by its function values.

As the former name *Evolving Algebras* suggests, states may *evolve* over time. That is, function values may change from one state to the next. We get a simple notion of behaviour in terms of state sequences

$$A_0 \xrightarrow{\delta_1} A_1 \xrightarrow{\delta_2} A_2 \xrightarrow{\delta_3} \ldots$$

We call such a sequence a *run* of the (abstract state) machine with A_i denoting possible states and δ_i the transitions between states.

To specify precisely the way a system may behave, i.e., its possible transitions, we use the notion of *transition rules* within our ASM framework. Transition rules specify possible state changes, i.e., transitions from one state to the next. Thus, a transition rule specifies a set of pairs (A_i, A_{i+1}). It restricts the predecessor state A_i implicitly through a certain precondition, which is called a *guard*. If the guard is satisfied in a state A_i, then the transition rule is applicable and will be *fired*. The successor state A_{i+1} is given in terms of changes on certain function values with respect to the predecessor A_i. That is, we denote the change of particular functions as *updates* and consider all other function values as unchanged (excluded are *external functions*, which may change arbitrarily to model non-determinism or inputs).

As ASM comprise states and state transitions, they can be viewed as a means of describing labelled transition systems or temporal structures (i.e., Kripke structures) in an operational manner.

The description of states by means of many-sorted structures or algebras is used in several specification methods. However, describing dynamics of a system by means of a set of transition rules is particular to ASM although we may find similar approaches in [GR97] and [CH85], where a notion for state transitions is attached to an algebraic method. In contrast to those approaches, ASM are much easier to understand and can be used by engineers that are not familiar with algebra and logic since the overall framework is kept much simpler.

In the following section, we present a very simple example in order to provide the reader with a basic understanding of ASM. Some remarks about methodology show how to use ASM for specifying more complex systems. The remainder of this chapter gives the precise definition of the ASM formalism and describes briefly the ASM Workbench as a tool framework for ASM in the end.

2.1.1   A Simple Example

The *dining philosophers* is a short, well-known example for a distributed system of communicating processes. The situation to be modelled is the following: *n* philosophers are sitting at a round table in front of a particular plate. Each of them wants sometimes to think and sometimes to eat spaghetti. The restricted resource in this situation is the set of forks: on the table are only as many forks as philosophers. These forks are placed besides the plates. For eating every philosopher takes the forks that are placed right and left to his or her plate. If two neighboured philosophers want to eat they have a conflict over the shared fork (which is placed left to the plate of one and right to plate of the other). Figure 2.1 sketches the situation for five philosophers.

Figure 2.1: The table with five dining philosophers

The dining philosophers are an example for a distributed system with a bounded resource that has to be shared by the contributing processes: the processes are the philosophers, the bounded resource is the available forks for eating. We can model the system of dining philosophers by means of a *multi-agent* ASM.

In multi-agent ASM, an *agent* describes a process, which is a philosopher in the given example. Since every philosopher behaves in the same way, it is suitable to model the behaviour generally by means of a set of *transition rules* which are parameterised (see below, Figure 2.3). We use a variable *Self* as the parameter. In our example, *Self* ranges over all philosophers.

Figure 2.2 shows the state space of our ASM model. The entities that may occur are *philosophers*, *forks*, and the *status* of forks. We introduce three domains, where the domains of philosophers and forks are parameterised by the same constant value *maxPhil* indicating the number of contributing entities (e.g., five in Figure 2.1). The domain **BOOLEAN** of Boolean values is implicitly included in every ASM specification.

Additionally, we introduce functions for modelling the situation as shown in Figure 2.2. Some functions are static, i.e., these functions do not change their values. For instance, each philosopher is associated with a particular *left fork* and a particular *right fork* which cannot be changed. We model the association between philosophers and forks as a mapping.

static function **maxPhil** = 5

PHILOSOPHER $= \{phil_1, phil_2, \ldots, phil_{maxPhil}\}$
FORK $\qquad = \{fork_1, fork_2, \ldots, fork_{maxPhil}\}$
STATUS $\qquad = \{taken, released\}$

static function **left_fork** : PHILOSOPHER $\rightarrow$ FORK
$\qquad\qquad\qquad = \{phil_i \mapsto fork_i \mid 1 \leq i \leq maxPhil\}$

static function **right_fork** : PHILOSOPHER $\rightarrow$ FORK
$\qquad\qquad\qquad = \{phil_i \mapsto fork_j \mid 1 \leq i \leq maxPhil,$
$\qquad\qquad\qquad\qquad\qquad j = (i+1) \bmod maxPhil\}$

dynamic function **status** : FORK $\rightarrow$ STATUS
$\qquad\qquad\qquad$ initially $\{status(f) \mapsto released \mid f \in$ FORK$\}$
dynamic function **eating** : PHILOSOPHER $\rightarrow$ BOOLEAN
$\qquad\qquad\qquad$ initially $\{eating(p) \mapsto false \mid p \in$ PHILOSOPHER$\}$

external function **hungry** : PHILOSOPHER $\rightarrow$ BOOLEAN

dependent function **forks_free**($Self$) : BOOLEAN
$\qquad\qquad\qquad\qquad = status(\text{left\_fork}(Self)) = released$
$\qquad\qquad\qquad\qquad \wedge status(\text{right\_fork}(Self)) = released$

Figure 2.2: Dining philosophers as ASM: domains and functions

Others functions are dynamic and can change their value during a run. For instance, a philosopher may take a fork and start to eat or may release a fork and stop eating. Thus, he or she will change the status of the corresponding forks and will be eating or not eating. For the two corresponding dynamic functions *status* and *eating* we define an initial mapping: *status* is initially *released* for each fork, and *eating* is initially *false* for each philosopher.

Functions may also be external, i.e., they are a kind of oracle and their values are determined by the outside world. For instance, whether a philosopher is *hungry* or not is not controlled by our model.

A dependent function can be used as an abbreviation for a larger expression. In our example, we use the functions *forks_free(Self)* to abbreviate a boolean expression for each philosopher that indicates if their left and right forks are free.

Based on the specified domains and functions, we can now describe the way a philosopher behaves. The behaviour of each philosopher is given by a collection of parameterised transition rules. The rules for our example are shown in Figure 2.3. Whenever a philosopher feels hungry, he or she will check the status of his or her left and right fork. If they are free then he or she can start to eat (first rule). If the philosopher is not hungry anymore but still eating (i.e., he or she has taken both forks) then the philosopher stops eating and should release the forks.

Note that in our model, the notion of the status of forks has to be a *global* function, i.e., each philosopher can read and update a fork status. Otherwise a philosopher could not check if the predicate *forks_free* is true or not, and could not change the *status* of a fork. In contrast to the function *status*, all other functions are parameterised with the variable *Self* and are thus *local* for each philosopher. Local functions are accessible only for the corresponding philosopher that is determined by the value of *Self*. The notion of global functions contradicts the ideal of real distributed processes. For modelling distributed philosophers that cannot share

---

if hungry(Self)
    then if forks_free(Self)
        then status(left_fork(Self)) := *taken*
             status(right_fork(Self)) := *taken*
             eating(Self) := *true*

if ¬ hungry(Self)
    then if eating(Self)
        then eating(Self) := *false*
             status(left_fork(Self)) := *released*
             status(right_fork(Self)) := *released*

Figure 2.3: Dining philosophers as ASM: the transition rules

functions, the ASM model needs to be refined. This refinement is, however, omitted here.

A run of our system can be simulated by firing both transition rules simultaneously in the current state (starting with the initial state) until no state transition is possible. Whenever the guards *hungry* and *forks_free*, or *¬hungry* and *eating* are true in a state, the three updates of the corresponding rule will change the function values in one atomic step (i.e., simultaneously) and thereby define the next state. In our example, the system runs forever since the values of external functions can always change, and thus there is no final state. Generally, a run of an ASM is defined over *state changes* that are caused by *firing updates*. The mere testing of guards is not considered as an action.

In the next section, we explain the methodology of how ASM should be used for modelling complex systems. In particular, we give a notion for modelling on different *levels of abstraction*. This technique involves a stepwise refinement of the model until the level of programming code is reached.

2.1.2 General Practice for Modelling in ASM

The small example in the last subsection provides a first impression of how ASM models look. In this subsection, we point out some practical aspects of the ASM approach in general.

ASM can be seen as a general purpose notation. It is suitable for various kinds of systems in terms of the problem to be solved and also in terms of size (the broad range in the literature shows this, see [Hug]). The ASM language is given in terms of basic mathematic concepts: sets, functions, and some simple rule constructors. Of course, this basic language can be extended for special purposes, however, the intention is to keep it small. Also no proof system or calculus is prescribed.

Two major guidelines for using ASM as a specification language are mentioned here as they have some impact on our work.

1. Model on a natural level of abstraction.
   This advice recommends that the user focus on important parts of the system that is to be modelled. The model should be as simple as possible for documenting well the speciality of the problem at hand. As a consequence, less important details may be abstracted in the model for the benefit of conciseness and readability. The influence of these abstracted details on the overall system can be stated in a mathematical way by means of (formal or non-formal) assumptions. Abstracting means introducing non-determinism into the model.

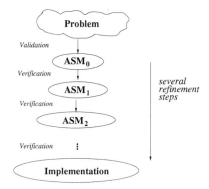

Figure 2.4: Different levels of abstraction

In ASM, non-determinism can be introduced in two ways: using external functions, which serve as oracles in that their value is determined through the outer world rather than the system, or by using the choose rule, which specifies that a parameterised transition rule is fired for a non-deterministically chosen instance of the parameter.

2. Prove the correctness of the system model in several steps.
   Proving correctness of a system implementation against its specification, the system model, usually addresses several features of the system. These features provide a natural structure for the proof: We model the system on different levels of abstraction, where each level is an extension or refinement of the next upper level by means of an additional feature that is specified. We break up the correctness proof into smaller steps by proving correctness between models on each two adjacent levels of abstraction.

Both guidelines yield a set of ASM that are modelled on various levels of abstraction: on the topmost level, the model is close to the (often informal) description of the system; the lower levels are more concrete, provide more and more information about a possible realisation and may lead to an implementation of the system. Figure 2.4 gives a sketch. The correctness of a resulting implementation can be *verified* against the more abstract ASM models: Considering two adjacent models, it has to be shown that the more concrete model "implements" the more abstract model. The most abstract model on the topmost level cannot be proved to be correct if the system description is informal. At this level we have to *validate* the model. To support this validation, the model should be modelled in a succinct and understandable way.

The notion of refinement that is used in the context of other specification languages is comparable to this approach. However, the ASM approach does not yet provide a suitable set of refinement rules such as that introduced in [Mor94].

2.2 The Language of Abstract State Machines

After intuitively introducing the formalism and the general practice of ASM, we now describe the theoretical framework, its syntax and semantics, in more detail.

We will follow the description given in [Gur95], [Gur97], [Bör95] and [Bör99], as well as [LEW96] (as far as common algebra is considered). This section is subdivided into definitions of states, locations, updates, and transition rules. The semantics of ASM is based on transition rules. Its definition is given in the end of this section.

2.2.1   States

The states of ASM are given as many-sorted first-order structures (or many-sorted algebras since we do not use relations in our signature). A structure is given with respect to a signature, logical names, variables and terms. To these we add an interpretation and a variable assignment. This allows evaluation of syntactical entities to semantical values.

Signature, logical names, variables and terms

Firstly, we define the base of structures: *signature, logical names, variables,* and *terms.*

Definition 2.1:   Signature and Logical Names

A signature Σ is given as a pair (S, Ω), where $S = (S_i)_{i \in I}$ denotes a set of *sorts* and Ω is a finite collection of function symbols, each with a fixed *arity*, i.e.,

$$f : S_1 \times S_2 \times \ldots S_k \to S_n$$

where S_i ($1 \leq i \leq k$) and S_n are sorts of the set of sorts S. We call $S_1, \ldots, S_k$ the *argument sorts* and S_n the *target sort* of the k-ary function. If $k = 0$ then f is a nullary function name.

Logical names are semantically related to truth values. They comprise the equality sign $=$, the constants *true, false* and *undef*, and the sort *Boolean* with its usual operators $\neg$, $\vee$, $\wedge$.

Sorts in a signature are names; these names will be used for addressing sets (see the definition of an algebra below). In the example of the last section (see Figure 2.2 on page 9) we use the names PHILOSOPHER, FORK, and STATUS as sorts. They are (in Figure 2.2) associated with sets. The function symbols are **maxPhil, hungry, left_fork, right_fork, status, eating,** and **forks_free.**

In [Gur97], the notion of *vocabulary* is used instead of a signature. A vocabulary defines the same mathematical concept as a signature in a slightly different way: A vocabulary γ has a finite set of function names corresponding to Ω. Instead of sorts S, however, *relational* function names are used for describing sets by means of characteristic functions. Moreover, the logical names are defined as being part of the vocabulary.

In this work, we base our definitions on the notion of signature for two reasons: First, having the notion of sorts (rather than characteristic functions) seems to be easier to understand. Thinking in terms of sorts works toward the typed version of ASM, ASM-SL, the language we are using within our tool-framework ASM Workbench (see Section 2.3). Secondly, it coincides with the given notion of first-order logic within the MDG framework (see chapter 4.2). Defining syntax and semantics with respect to sorts eases the definition of the MDG representation. The price we have to pay is that we have to refer to sorts whenever functions are involved. For our concerns, however, this seems to be reasonable to pay.

As common in algebraic specification languages, we use *variables* and *terms* over the signature as mathematical objects of our structure.

In an ASM specification a variable can occur for instance in the definition of a dependent function, or a universe (defined as a set comprehension), or in the transition rules. They are a means for parameterisation.

Definition 2.2: Variables

We associate with a signature $\Sigma = (S, \Omega)$ sets of variables of sort S_i, denoted by V_i. The entire set of used variables is $V = (V_i)_{i \in I}$.

We suppose that each V_i is sufficiently large to always provide "fresh" variables that are different from all others, if necessary. With the definition of variables we can define the notion of terms over a signature Σ.

Definition 2.3: Terms

Let $\Sigma = (S, \Omega)$ be a signature and V a set of variables for Σ. Terms over all sorts in S, denoted by $T_{\Sigma(V)} = (T_{\Sigma(V),S_i})_{s_i \in S}$ are now defined recursively, as in first-order logic

- every variable is a term, $V \subseteq T_{\Sigma(V)}$
- if $f \in \Omega$ is a function symbol with arity $f : S_1 \times \ldots S_k \to S_n$ and $t_1, \ldots t_k$ are terms of corresponding sorts, i.e., $t_1 : S_1, \ldots, t_k : S_k$ then $f(t_1, \ldots, t_k)$ is a term.

A term of sort Boolean is referred to as a *boolean term*. We extend the definition of terms by *first-order terms*.

- if $v : V_i$ is a variable, and $g(v)$ and $s(v)$ are boolean terms then

$$(\forall v : g(v))\, s(v)$$
$$\text{and} \quad (\exists v : g(v))\, s(v)$$

are terms, with *head variable v, guard $g(v)$* and *body $s(v)$*. As usual we say that v is a *bound variable* in the term.

The set of variables occurring in a term t is denoted by $Var(t)$. A *ground term* denotes a term containing no variables, i.e., $Var(t) = \emptyset$.

States

Now we define the notion of *states* of an ASM in terms of a many-sorted algebra for Σ (following [Gur97]). We start with the definition of the latter.

Definition 2.4: Algebra

A many sorted *algebra* for the signature $\Sigma = (S, \Omega)$ is given by

- *(carrier) sets* S_i^A for each sort $S_i \in S$
- an interpretation for each function symbol $f : S_1 \times S_2 \times \ldots \times S_k \to S_n$ in Ω given by a total function

$$f^A : S_1^A \times S_2^A \times \ldots \times S_k^A \to S_n^A$$

where $S_1^A \times S_2^A \times \ldots \times S_k^A$ is called the domain of the function, $dom(f)$, and S_n^A the range of the function, $ran(f)$.

In an algebra, we interpret the logical names as distinguished elements. We do not distinguish between the names and the truth values *true* and *false*, and = is identified with the equality on a carrier set. The sort *Boolean* is interpreted as the set of truth values.

Definition 2.5:   State

A *state* A is a many sorted algebra as it is defined in (2.4).

In the context of ASM, a carrier set S_i^A of a states A is called a *universe*. We denote with S^A the *super-universe* of a state. It contains the universes S_i^A and also new elements that are not contained in any of the universes; these new elements comprise the reserve (see below); that is, $S^A \supset \bigcup_i S_i^A$.

A state is implicitly given in an ASM model; it depends on the evaluation of functions which can change from one state to another. For example, in the dining philosophers (see Figure 2.2) we associate the sort **PHILOSOPHER** with the universe $\{phil_1, \ldots, phil_{maxPhil}\}$. The functions are declared as mappings over the universes (e.g., **eating** : **PHILOSOPHER** $\rightarrow$ **BOOLEAN**). A particular state of the dining philosophers is given as the current evaluation of all functions (e.g., **eating** can be *true* in the considered state).

As states are mostly described in terms of functions, we distinguish different kind of functions.

- *Dynamic* functions may change their interpretation during a run.

- *Static* functions have a fixed interpretation, they may be considered as constants (of arbitrary but finite arity).

- *External* functions (or oracles) are a sort of dynamic functions but their interpretation cannot be changed by the system itself but rather by the outer world (the name "oracle" indicates this difference quite well). They are a perfect means to model reactive systems which are influenced by input or environmental behaviour (as for instance, sensor values to be read by the system).

- *Dependent* functions are defined in terms of other functions given in our structure. They can be seen as macros in order to shorten expressions.

- Relational functions are functions with target sort Boolean. Elsewhere denoted as characteristic functions, they are a means to describe sets. Syntactically, however, relational functions are not distinguished from others.

Assignments and evaluation

Given an assignment for the variables, we can evaluate each term to its current value. Variable assignment, function interpretation (as already given through the notion of structure) and term evaluation forge links between syntax and semantics, i.e., between structure and logics.

To provide terms with semantics, we need to assign values to variables first. This is done by a mapping called *variable assignment*.

Definition 2.6:   Variable Assignment

A variable assignment over a state A is a function $\zeta : V \rightarrow S^A$ such that variables of a particular sort $S_i \in S$ are mapped to the corresponding universe, $\zeta : V_i \rightarrow S_i^A$.

Accordingly, a boolean variable $v \in V_j$ (where $S_j = Boolen$) is mapped into Boolean $\zeta(v) \in \{true, false\}$. Given a variable assignment we can define an expanded state.

Definition 2.7:   Expanded State

Given a variable assignment ζ over state A, we denote the pair $B = (A, \zeta)$ as the *expanded state*. If ζ assigns the value $a \in S^A$ to the variable v in a state A we denote the partially expanded state as

$$B(v \mapsto a) = (A, \zeta(v \mapsto a)).$$

A state A and a term t are *appropriate* for each other if the signature of A contains all function symbols occurring in t and $Var(t)$ is included in the set of variables of A.

Definition 2.8:   Term Evaluation

Let $B = (A, \zeta)$ be an expanded state and t range over terms appropriate for state A, then the value $Val_B(t)$ of t at state A is defined as follows

- if the term is a variable $t = v$, then $Val_B(t) = \zeta(v)$
- if $t = f(s_1, \ldots, s_k)$ then its value is given by induction over the term structure

$$Val_B(t) = f^A(Val_B(s_1), \ldots, Val_B(s_k))$$

If the term t is a ground term, then $Val_B(t) = Val_A(t)$.

The evaluation of first-order terms depends on the notion of *satisfiability* on boolean terms. As logical names are given in algebra A and the universe Boolean is identified with the logical truth values, we can speak about evaluation of boolean terms. If a boolean term evaluates to *true* in an expanded state B, then we say t holds at B or B *satisfies* t, denoted as $B \models t$.

The guard $g(v)$ of a first-order term is a boolean term. Assume $v : V_i$ then we define the *range* of the guard as the set of satisfying variable assignments:

Definition 2.9:   Range

$$Range_B(v : g(v)) = \{a \in S_i^A \mid B \models g(a)\}$$

where $g(a)$ denotes the term g in which each occurrence of variable v is substituted by value a.

Now the evaluation of first-order terms can be defined in the following way:

Definition 2.10:   Evaluation of first-order terms

$$Val_B\big((\forall v : g(v))\, s(v)\big) = \begin{cases} true & \text{if } Val_{B(v \mapsto a)}(s(v)) = true \\ & \text{for all } a \in Range_B(v : g(v)) \\ false & \text{otherwise} \end{cases}$$

$$Val_B\big((\exists v : g(v))\, s(v)\big) = \begin{cases} true & \text{if } Val_{B(v \mapsto a)}(s(v)) = true \\ & \text{for some } a \in Range_B(v : g(v)) \\ false & \text{otherwise} \end{cases}$$

Reserve

Some systems or programs introduce new elements during a computation. For instance, new memory space may be allocated by a program. In order to provide a state with new or fresh elements, we introduce the notion of the *reserve* of a state.

15

The reserve is infinite and contains all elements of S^A that are not contained in any of the universes S_i^A. Every function that is defined over A and its universes $f : S_1 \times \ldots \times S_k \to S_n$ will accordingly evaluate to *undef* if any of its arguments belongs to the reserve. No function may output an element of the reserve.

For the variable assignment ζ, we require that it does not map the same reserve element r to different variables: $\zeta(v_1) = \zeta(v_2) = r \Rightarrow v_1 = v_2$.

We call an entity a of A *non-reserve* if $a \in S_i^A$ for any universe $S_i^A \subset S^A$.

2.2.2 Locations and Updates

In ASM, states are described through functions and their current interpretation. A state may change and transit to the next state by changing its function values locally, i.e., at particular points. In order to capture this notion, we use *locations* and *updates* of locations.

Definition 2.11:   Location and its Value

> A *location* of a state A is a pair $loc = (f, \overline{a})$ where f is a symbol for a dynamic function in Ω with $f : S_1 \times \ldots \times S_k \to S_n$ and $\overline{a}$ is a k-tuple of elements of S^A with $\overline{a} : S_1^A \times \ldots \times S_k^A$. An element $f^A(\overline{a}) : S_n$ is the *value* of the location loc in the state A.

In a state A, each term $f(t_1, \ldots, t_k)$, where f is dynamic, is associated with a location $\big(f, \big(Val_A(t_1), \ldots, Val_A(t_k)\big)\big)$. That is, to determine the value of a location we first have to evaluate the argument terms t_i ($1 \leq i \leq k$) to values $Val_A(t_i) = a_i \in S_i^A$. The resulting location $(f, (a_1, \ldots, a_k))$ is mapped to the value $f^A(Val_A(t_n)) = f^A(a_n)$. Thus, we can describe any state by mappings of locations to their (current) values in A, i.e., the current interpretation of f at $\overline{a}$ is denoted as $\{(f, \overline{a}) \mapsto f^A(\overline{a})\}$. That is, $(f, \overline{a})$ equals semantically $f^A(\overline{a})$ in state A.

For example, the model of the dining philosophers, given in the last subsection, contains among others the locations **status**($fork_1$), **status**($fork_2$), **eating**($phil_1$), **eating**($phil_2$).

For changing values of locations we use the notion of an update.

Definition 2.12:   Updates

> An *update* of state A is a pair $\alpha = (loc, val)$ where loc is a location of sort S_n and val, denoting the value, is an element of S_n^A or *undef*. To *fire* an update α at A, we set val to be the new value of location loc and redefine f to map loc to val. This redefinition causes A to be changed. The resulting state A' is a successor state of A with respect to α. All other locations in A' are unaffected and keep their value as in A.

Within a step from one state to the next state we can fire several updates. Therefore we say more generally within a step from state A to A' an *update set* is fired at A. Firing an update set Δ is firing each of its members $\delta \in \Delta$ *simultaneously*.

If we consider several updates to be fired at the same time, these updates need to be *consistent*, that is, they should not affect the same location in a different way.

Definition 2.13:   Consistency

> For locations loc_i and loc_j in a state A, and the values val_i and val_j in S^A we say the updates $\alpha_i = (loc_i, val_i)$ and $\alpha_j = (loc_j, val_j)$ are *consistent* if and only if $(loc_i = loc_j) \Rightarrow (val_i = val_j)$. An update set Δ is consistent if no inconsistencies between its members occur.

If we fire an inconsistent update set at A, it is defined in [Gur97] that none of the updates should be fired and the state remains unchanged. Obviously, we cannot assign two different values to the same location simultaneously.

Update sets are unambiguously determined by their function symbols and non-reserve elements. Thus, we do not distinguish update sets which differ only in the names of the contributing reserve elements. Two update sets are said to be *equivalent* (and thus indistinguishable) if either both are inconsistent updates sets or both are consistent and there is an automorphism π on the reserve elements that maps one update set onto the other. ([Gur97] introduces equivalence classes of update sets, called actions.)

2.2.3   Transition Rules

Transition rules are a means of specifying the behaviour of our system, i.e., the possible changes from one state to the next. The syntax is given by atomic rules for updating and rule constructors that allow for more complex specifications. The semantics is described denotationally as introduced in the second part of this section.

Syntax of Transition Rules

The **skip rule** is the simplest transition rule. It is denoted as

$$\texttt{skip} \qquad\qquad (2.1)$$

This rule specifies an "empty step". No function value is changed.

The **update rule** plays a central role for "evolving" states. It is an atomic rule denoted as

$$f(t_1,\ldots,t_n) \; := \; t \qquad\qquad (2.2)$$

It describes the update, i.e., the change of interpretation of function f at the place given by $(t_1,\ldots,t_n)$ to the current value of t. Function f given by signature Σ is called the *head* of the update rule with arity n. For $1 \leq i \leq n$, t_i and t are elements of $T_{\Sigma(V)}$ and denote terms of the signature.

A **conditional rule** is the most common means of specifying a precondition for updating.

$$
\begin{aligned}
&\texttt{if } g\\
&\quad \texttt{then } R_1\\
&\quad \texttt{else } R_2\\
&\texttt{endif}
\end{aligned}
\qquad\qquad (2.3)
$$

The *guard* g is a boolean term in $T_{\Sigma(V)}$, which we assume is decidable. R_1 and R_2 denote arbitrary transition rules. If the guard evaluates to *true* then, obviously, R_1 fires at the current state. Otherwise, choosing the else-case, R_2 is applied. We may omit the else case in a conditional rule if R_2 is a skip rule.

A **block rule** groups a set of transition rules

$$
\begin{aligned}
&\texttt{block}\\
&\quad R_1\\
&\quad R_2\\
&\quad \ldots\\
&\texttt{endblock}
\end{aligned}
\qquad\qquad (2.4)
$$

The transition rules R_i for $i \geq 1$ are to be fired simultaneously. With a block rule we construct the overall ASM *program*. All transition rules that specify the behaviour of the machine are grouped into a block indicating that all of them are fired simultaneously at each step.

The **do-forall rule** is a generalisation of a block rule

$$\begin{array}{l} \texttt{do forall } v \,:\, g(v) \\ \qquad R_0(v) \\ \texttt{enddo} \end{array} \qquad\qquad (2.5)$$

The *head variable* v determines the valuation of the boolean term $g(v)$, the *guard* of the rule. If $g(v \mapsto a)$ is satisfied for some value a in the current state then transition rule R_0 depending on a fires. This constructor can be seen as a parametrisation of conditional rules. It allows transition rules that have to be applied to a number of elements to be shortened.

The **choose rule** is a rule for modelling non-determinism.

$$\begin{array}{l} \texttt{choose } v \,:\, g(v) \\ \qquad R_0(v) \\ \texttt{endchoose} \end{array} \qquad\qquad (2.6)$$

The rule R_0 is applied for some arbitrary value a that satisfies the guard $g(v)$ in the current state. The choice of the value among several possible elements is non-deterministic. If no element satisfies the condition, the application of the choose rule has no impact; it is an "empty step".

The **import rule** is a means of extending the signature of our state.

$$\begin{array}{l} \texttt{import } v \\ \qquad R_0(v) \\ \texttt{endimport} \end{array} \qquad\qquad (2.7)$$

The variable v is assigned a "fresh" or new element that does not occur in any of the sorts of the signature. (That is, the element is taken from the reserve, see Section 2.2.1 on page 15.)

Note that without using an import rule, the state space of an ASM, given by all universes, is fixed.

Semantics of Transition Rules

The semantics of a transition rule is given as a *denotation*. Since it depends on the evaluation of terms at a current state, a denotation is a function that maps a transition rule and an expanded state to an appropriate *set of updates* (see Definition 2.12 on page 16). For any transition rule R and expanded state $B = (A, \zeta)$ the denotation of R on state A and variable assignment ζ is an update set denoted as $\Delta_B(R)$.

We give a definition for $\Delta_B(R)$ for each of the rule constructors introduced syntactically above.

Skip rule: The "empty step" affects no updates, we get the empty set.

$$\Delta_B(\texttt{skip}) \;=\; \{\} \qquad\qquad (2.8)$$

Update rule: Updating a single location is formalised as a singleton:

$$\text{for any } R \equiv f(t_1, \ldots, t_n) \; := \; t$$

$$\Delta_B(R) \; = \; \{\alpha\} \tag{2.9}$$

where $\alpha = (loc, \; val)$ is an update,
with $loc = (f, (Val_B(t_1), \ldots, Val_B(t_n)))$ denoting the *location*
and $val = Val_B(t)$ its new *value* in expanded state B.

Conditional rule: The update set of a conditional rule depends on the evaluation of the guard in the current state:

$$\text{for any } R \equiv \texttt{if } g \texttt{ then } R_1 \texttt{ else } R_2 \texttt{ endif}$$

$$\Delta_B(R) \; = \; \left\{ \begin{array}{ll} \Delta_B(R_1) & \text{if } B \models g \\ \Delta_B(R_2) & \text{otherwise} \end{array} \right. \tag{2.10}$$

Block rule: Within the block rule we are gathering sub-rules that have to be fired simultaneously. This is formalised as a union of the update sets corresponding to the sub-rules:

$$\text{for any } R \equiv \texttt{block } R_1 \; \ldots \; R_n \texttt{ endblock}$$

$$\Delta_B(R) \; = \; \Delta_B(R_1) \cup \ldots \cup \Delta_B(R_n) \tag{2.11}$$

Do-forall rule: Assume that $R_0(a)$ is a shorthand for rule $R_0(v)$ after substituting all occurrences of variable v by the element a. The denotation of a do-forall rule is given as the union of update sets for all sub-rules $R_0(a)$ for any value a such that $g(a)$ is satisfied in the current state B. Accordingly it is said we apply $R_0(a)$ for each value a that satisfies $g(a)$ in the current state. With $\mathcal{RG} = Range_B(v : g(v))$, and $B(v \mapsto a) = (A, \zeta(v \mapsto a))$ we get the following denotation:

$$\text{for any } R \equiv \texttt{do forall } v \; : \; g(v) \;\; R_0(v) \texttt{ enddo}$$

$$\Delta_B(R) \; = \; \bigcup_{a \in \mathcal{RG}} \Delta_{B(v \mapsto a)}(R_0(v)) \tag{2.12}$$

Import rule: The semantics of an import rule with body R_0 is given by the update set corresponding to $R_0(v)$ after substituting every occurrence of variable v by some fresh element a from reserve (i.e., $S^A \backslash \bigcup_i S_i^A$). Let $B' = (A', \zeta)$, and correspondingly $B'(v \mapsto a) = (A', \zeta(v \mapsto a))$, be the expanded state of a state A', where variable v is of sort S_i, i.e., $v \in V_i$, and A' is the state we get from A by enlarging one of the universes S_i^A with a fresh element from reserve, i.e., $S_i^{A'} = S_i^A \cup \{a\}$. This formalised as follows:

$$\text{for any } R \equiv \texttt{import } v \;\; R_0(v) \texttt{ endimport}$$

$$\Delta_{B'}(R) \; = \; \Delta_{B'(v \mapsto a)}(R_0(v)) \tag{2.13}$$

$$\text{for some } a \in S^A \backslash \textstyle\bigcup_i S_i^A$$

Since we do not distinguish update sets which differ only in the contributing reserve elements (see Section 2.2.2 for the notion of equivalence of update sets) the import rule is not considered to provide non-determinism. It makes no difference which of the reserve elements a is taken to be imported.

In order to define the semantics for the choose rule we have to introduce a notion that enables us to capture non-deterministic behaviour. Generally, non-determinism can be formalised by a set of denotations rather then a single (thus determined) denotation. If a set of update sets gives the denotational semantics of an ASM then we can state that each member of this set describes a possible denotation. In order to provide a common framework, we introduce a non-deterministic semantics for each rule constructor, and add the denotation for the choose rule.

Non-deterministic Semantics of Transition Rules

The non-deterministic semantics for transition rules is given in terms of sets of denotations, that is *sets of update sets*, denoted as $\mathcal{N}_B(R)$ for an expanded state $B = (A, \zeta)$ and rule R. They are defined for each rule constructor.

Skip rule, update rule: The semantics of skip and update rules is deterministic and thus a singleton:

$$\text{for any } R \equiv \texttt{skip} \text{ or } R \equiv f(t_1, \ldots, t_n) := t$$

$$\mathcal{N}_B(R) = \{\Delta_B(R)\} \tag{2.14}$$

Conditional rule: The non-deterministic semantics of conditional rule is just reducible to the semantics of its sub-rules:

$$\text{for any } R \equiv \texttt{if } g \texttt{ then } R_1 \texttt{ else } R_2 \texttt{ endif}$$

$$\mathcal{N}_B(R) = \begin{cases} \mathcal{N}_B(R_1) & \text{if } B \models g \\ \mathcal{N}_B(R_2) & \text{otherwise} \end{cases} \tag{2.15}$$

Block rule: The semantics of block rules is the set of unions of any members of the non-deterministic denotations of the sub-rules:

$$\text{for any } R \equiv \texttt{block } R_1 \ \ldots \ R_n \texttt{ endblock}$$

$$\mathcal{N}_B(R) = \{\Delta_B(R_1) \cup \ldots \cup \Delta_B(R_n) \mid \\ \forall i. \ 1 \leq i \leq n, \ \Delta_B(R_i) \in \mathcal{N}_B(R_i)\} \tag{2.16}$$

Do-forall rule: The non-deterministic semantics of the do-forall rule is similarly given by the set of unions of the non-deterministic denotations of the sub-rules. We use $R_0(a)$ as a shorthand for rule $R_0(v)$ after substituting all occurrences of variable v by the element a and $\mathcal{RG}$ for $Range_B(v : g(v))$. The non-deterministic semantics is given as follows:

$$\text{for any } R \equiv \texttt{do forall } v \ : \ g(v) \ R_0(v) \texttt{ enddo}$$

$$\mathcal{N}_B(R) = \{setTuple(t) \mid t \in \mathcal{N}_B(R_0(a_1)) \times \ldots \times \mathcal{N}_B(R_0(a_n))\} \tag{2.17}$$

where $\{a_1, \ldots, a_n\} = Range_B(v : g(v))$, and $setTuple(s_1, \ldots, s_n) = s_1 \cup \ldots \cup s_n$, i.e., any tuple $(s_1, \ldots, s_n)$ of sets s_i is mapped by function $setTuple$ to the union of the sets s_i.

Import rule: The non-deterministic semantics of the import rule is similar to the deterministic semantics. The import rule does not introduce non-determinism by itself since update sets which differ only in the contributing reserve elements are considered as equivalent (see Section 2.2.2). Only the sub-rule R_0 may have a non-deterministic semantics.

for any $R \equiv$ import v $R_0(v)$ endimport

$$\mathcal{N}_{B'}(R) \; = \; \mathcal{N}_{B'(v \mapsto a)}(R_0(v)) \qquad\qquad (2.18)$$

for some $a \in S^A \backslash \bigcup_i S_i^A$

Choose rule: The choose rule demands the non-deterministic choice of an element $a \in A$ that satisfies guard $g(v)$. After replacing all occurrences of variable v in the state by element a we apply the denotation of body R_0. $\mathcal{RG} = Range_B(v : g(v))$ is used as a shorthand:

for any $R \equiv$ choose $v \; : \; g(v) \; R_0(v)$ endchoose

$$\mathcal{N}_B(R) \; = \; \bigcup_{a \in \mathcal{RG}} \mathcal{N}_{B(v \mapsto a)}(R_0(v)) \qquad\qquad (2.19)$$

Every element of $\mathcal{N}_B(R)$ is a suitable denotation for R, i.e., a can be chosen *non-deterministically*. If $Range_B(v : g(v)) = \emptyset$ then $\mathcal{N}_B(R) \; = \; \{\emptyset\}$.

2.3   The ASM Workbench

The ASM Workbench (ASM-WB), developed by Giuseppe Del Castillo (see [Cas00]), provides a general framework for ASM tools. Due to the benefit of tool support, Del Castillo extended the ASM language with a type system. The resulting *ASM specification language* (ASM-SL) allows static analysis by means of type checking. The kernel of the tool environment comprises basic functionality for developing ASM-SL models: a parser for providing an internal representation of the model data in abstract syntax, a type-checker for checking type correctness, an interpreter for simulating and animating runs of ASM models. Figure 2.5 shows

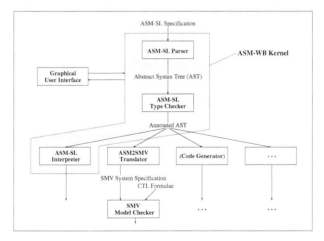

Figure 2.5: The ASM Workbench

the raw architecture of the tool environment[1]. The grey part depicts the kernel. Further basic functionality is given by a graphical interface for editing the model and representing states and state changes.

The ASM-WB is designed as an open tool environment which is extensible. For integrating other tools, *transformation algorithms* might be added that serve as interfaces. Figure 2.5 suggests an interface to the model checker SMV ([McM93]), called the "ASM2SMV Translator" here, or a possible interface for a code generator. The development of interfaces is supported by an exchange format given as *annotated abstract syntax trees* (AST). The AST format is a textual representation of the internal data representation that can be used as input for transformation tools. The ASM-WB parser and its type checker provide a front-end for editing and syntactically checking ASM models before they are treated by another external tool.

In this thesis, we use the ASM-WB as a core tool framework. The contribution of this thesis is two interfaces for supporting ASM with model checking tools: an interface to the SMV tool, and an interface to the MDG-Package ([CZS$^+$97]). The former was developed in cooperation with Del Castillo.

In order to give a background to this work, we describe the theoretical framework for model checking in the next chapter.

---

[1]This Figure originally appeared in [Cas00] and is reproduced here with the kind permission of the author.

Chapter 3

Model Checking: An Overview

3.1 Introduction

Generally, the verification of a program requires proving, in a rigorous or formal way, that the program satisfies a specification. That is, verification involves comparing two descriptions of what the program should do.

Traditionally, this is done by using logic as a formal framework for deductive reasoning, which is, when tool supported, also referred to as *theorem proving*. First, a general axiomatisation of the domain of interest (data structures of the program) within the logic has to be developed. The program needs to be incorporated (or translated) into the logic. Then we may try to prove the specification to be correct (with respect to the program) by means of applying the precisely given deduction rules of the axiomatic framework.

Model checking, in contrast to the traditional approach, is much simpler since it does not aim at being fully general. It was initially restricted to finite state systems and, although this limitation is no longer absolute, it is only applicable to system whose states and state transitions have short and simple descriptions. Typically, the intricacy of those systems resides more in the *control* than in the data as, for example, in hardware descriptions, control systems, or concurrent protocols. They are generally referred to as *reactive systems*. Their core functionality can be described by sequences of possible interactions with the environment.

Due to the restriction to simple and finite systems, model checking is fully algorithmic. It is an automatic machine-implemented approach that requires no user interaction. Moreover, model checking provides explicit information about possible erroneous behaviour of the system for which the checked property fails. Given this, model checking can be viewed as an excellent debugging tool that checks every possible behaviour.

3.2 The Overall Framework

Model checking has developed along two different major directions: the *logical approach* based on fixpoint computation, and the approach adopted from *automata theory* which is based on proving language containment.

In this work we make use of fixpoint-based algorithms. Representative tools for this direction are SMV ([McM93]), VIS ([Gro96]), SVE ([F$^+$96]), and the MDG-Package ([CZS$^+$97]). We will start with an introduction of this logical approach.

Since some well known tools implement the algorithms of the automata theoretical approach (e.g., SPIN ([Hol97]), COSPAN ([HHK96]), and FDR ([For96, Ros94])[1]), we will briefly explain the basic ideas behind this approach as well. This is done in Subsection 3.2.2. The remaining part of this chapter, however, deals with the logical approach based on fixpoint computations.

3.2.1 The Logical Approach

What we call the *logical approach* here comprises the algorithms based on fixpoint computation. The requirements in this framework are given as temporal logic formulas. In the following, the notion of structures and temporal logics is introduced, and we formalise the model checking problem in most general terms.

The Underlying Principle

In general, a model checker checks that a given logical formula is satisfied on a specific finite domain. That means, we are *checking* that the domain is a *model* of the formula.

The finite domain is a program or a finite-state transition system. The logic needs to provide formulas that can be directly evaluated on the transition system. This is the case for *propositional temporal logic*. In classical propositional logic, the truth value of a formula is computed according to the truth value of the involved propositions, which are boolean variables. In addition to that, the world in temporal logics has several possible *states*. Truth values of propositions are given with respect to the state, and they may change as we move from one state to the next. Also *temporal operators* are available that can express properties with respect to some or all future states. We can identify these states with the *system states* (of the finite domain) and the moves with the system's *state transitions*. This way, progress properties of a given system, such as safety and liveness, can be directly represented in propositional temporal logic. Checking that the system satisfies a given temporal property involves simply evaluating the temporal formula on the transition system.

The Transition System

The transition system to be model checked is non-deterministic. The non-determinism may either be caused by an interleaving semantics of concurrent transitions or by absence of information about the behaviour of a system component or the environment. If the system has inputs, all possible values of these inputs are checked during the model checking process.

Most commonly, a transition system is defined as a *temporal structure* (or Kripke structure). A temporal structure is given by a set of states, a transition relation defined as pairs of states, and a function labelling each state with a set of propositions.

Definition 3.14: Temporal Structure

A *temporal structure* $M = (S, S_0, R, L)$ comprises

i.) a set of *states* S

ii.) a set of *initial states* S_0, where $S_0 \subseteq S$

iii.) a *transition relation* $R \subseteq S \times S$,
where $\forall s \in S. \exists s' \in S. (s, s') \in R$, i.e. R is total

---

[1] The model checker FDR ([For96], [Ros94]) checks CSP processes by proving failures/divergence refinement. Whilst the CSP processes are compiled into automata, the refinement relation turns into a language containment relation on these automata.

iv.) a *label function* $L : S \to \mathcal{P}(\mathcal{V})$,
   where $\mathcal{V}$ is the set of atomic propositions
   (i.e., boolean variables expressing what is observable in a state).

M is a *finite* transition system if the set of states S is a finite set. A *path* of a structure M starting at an initial state $s_0 \in S_0$ is an infinite sequence of states $\underline{s} = s_0 s_1 s_2 \dots$ such that $(s_i, s_{i+1}) \in R$ for all $i \geq 0$. With $\underline{s}^k$ (given any $k \geq 0$), we denote a *suffix* of a path $\underline{s}$ starting at state s_k, i.e., $\underline{s}^k = s_k s_{k+1} \dots$

Generally, a temporal structure formalises the notion of branching time behaviour. A model of the structure comprises many infinite paths which reflect the possibilities for the non-deterministic behaviour of the system. Unwinding these paths, the model can be depicted as an infinite and acyclic *computation tree*. A computation tree comprises nodes and edges. Each node is labelled by the label function L that characterises sets of states. That is, each node represents a set of states. The root node of a computation tree represents the set of initial states. According to the given transition relation R, each node is connected via edges with its successor nodes.

The Requirements

The property to be checked—also called the *(requirement) specification* —is a formula in a temporal logic (branching time temporal logic or linear time temporal logic) that specifies the required behaviour.

Such a temporal logic does not deal with time values explicitly or time intervals but rather with the notion of sequences of states which describe possible computations (or behaviour) of the system. Their semantics is given with respect to a temporal structure.

Since we want to treat non-deterministic systems (deterministic systems can be validated by driving a simple test) we are dealing with behaviour that may evolve in several different ways. That is, either each state may have multiple successors in terms of a branching time, or behaviour is given as a set of various paths evolving in linear time.

In *branching time* temporal logic multiplicity of behaviour can be specified explicitly by means of considering properties for *all* or *some* successor states. For *linear* temporal logic this is expressed implicitly. The logic is designed to describe linear sequences of states. A formula is satisfied in a linear temporal structure if all paths that are a model of the structure satisfy the formula.

The Model Checking Problem

Let M be the given temporal structure, $s \in S$ a state in M, and φ a temporal logic formula. Then a general formalisation of the *model checking problem* is introduced as

$$M, s \models \varphi. \tag{3.1}$$

If we identify s with an initial state $s_0 \in S_0$ we are *checking* if the system M is a *model* of the requirement φ.

The model checking approach depends on the logic used for the specification. Each approach requires its own algorithm. Within the logic of branching time, properties are related to sets of states. If we assume that S is a finite set then the corresponding domain $\mathcal{P}(\mathcal{S})$ (the power set of states) is a complete lattice. Thus, fixpoint computations can be used as efficient algorithms for a state space exploration in finite transition systems.

Formulas of a linear temporal logic are related to single paths. With a fixpoint computation, however, only sets of satisfying states may be computed, but not a path. Instead the notion of semantical tableaus is adapted for linear temporal logics. Formulas can be represented in terms of a tableau, which is a special temporal structure.

3.2.2   Model Checking with Automata

Within the automaton approach both the system model and the requirements are given as finite automaton on words (ω-automaton). That is, both descriptions are given in the same notation.

An ω-automaton is given as a five tuple $(\Sigma, S, \Delta, S^0, F)$, where Σ is a finite alphabet, S is the set of states, and $S^0 \subset S$ the set of initial states, $\Delta \subseteq S \times \Sigma \times \mathcal{P}(S)$ the (non-deterministic) transition relation, and F the set of final states (see also Chapter 4 in [vL90]).

Computations (or paths in the structure) can be given in terms of words over the alphabet, $w \in \Sigma^*$. Thus, we can model requirements to be checked as a word. A *run* over a word $w = a_1 a_2 \ldots$ is a sequence of states $\underline{s} = s_0 s_1 \ldots$ with $s_0 \in S_0$ and $s_i \in \Delta(s_{i-1}, a_i)$ for all $1 \leq i$.

Since most reactive systems are designed for endless execution, the computations have to be modelled as infinite paths. By replacing the set of final states F with a set of *accepting* states F' we are able to model infinite behaviour as well. An infinite run $\underline{s} = s_0 s_1 s_2 \ldots$ in an automaton A is an *accepting run* if there is some designated state that occurs infinitely often in the run. That is, for some $s \in F'$ there are infinitely many states s_i in $\underline{s}$ such that $s_i = s$. The infinite word w is accepted if there is an accepting run of A over w. The simplest automaton over infinite words are Büchi automaton (see for instance [VW86]). Note that a temporal structure as well as a temporal logic formula (over branching or linear time) can be transformed into Büchi automaton.

Once both models, system and requirements, are given as automata, A and A', solving the model checking problem can be done by checking language containment of the automaton. The set of possible behaviours of the system is given as the set of words accepted by the automaton, i.e., the language of the automaton $\mathcal{L}(A) \subseteq \Sigma^*$. The same holds for the requirement automaton A'. *Language containment*

$$\mathcal{L}(\mathcal{A}) \subseteq \mathcal{L}(\mathcal{A}')$$

formalises[2] that every possible behaviour of the system ($v \in \mathcal{L}(\mathcal{A})$) is suitable according to the requirements ($v \in \mathcal{L}(\mathcal{A}')$). That is, system A satisfies the requirements A'.

If $\overline{\mathcal{L}(\mathcal{A}')}$ denotes the complement of the required behaviour $\mathcal{L}(\mathcal{A}')$, then the model checking problem is given as

$$\mathcal{L}(\mathcal{A}) \cap \overline{\mathcal{L}(\mathcal{A}')} \ = \ \emptyset$$

We prove that there is no behaviour of A that is disallowed by A'. Following this approach, firstly the complement automaton $\overline{A'}$ has to be constructed that accepts $\overline{\mathcal{L}(\mathcal{A}')}$. In a second step, A and $\overline{A'}$ must be combined such that the resulting automaton accepts the intersection of $\mathcal{L}(\mathcal{A})$ and $\overline{\mathcal{L}(\mathcal{A}')}$. If this intersection is empty, then the specification A' holds for A. If the intersection is not empty, then the requirements are violated and the elements of the intersection provide counterexamples.

---

[2]Note that the language of an automaton is never empty for that it contains at least the empty word ε.

Based on this approach, Gerth, Peled, Vardi and Wolper ([GPVW95]) developed an algorithm for checking path formulas of linear temporal logic. If the requirement is given as a formula ψ, then the Büchi automaton for $\neg\psi$ is constructed. As an example, the model checker SPIN ([Hol97]) processes requirements that are specified as a temporal logic formula. The negated formula is transformed into the corresponding automata. SPIN also allows representing requirements as automaton that specifies the bad behaviour. It corresponds to the automaton $\overline{A'}$. This way the construction of the complement automaton can be avoided.

Within this work, the logical approach of fixpoint computation is used. We are interested in model checking logics for branching as well as linear time, further on called CTL model checking and LTL model checking, respectively. In Section 3.3 and Section 3.4, we introduce the logics and corresponding solutions for the model checking problem by means of efficient algorithms. In Section 3.5, we compare both logics by the complexity of the corresponding algorithms and their expressiveness.

3.3   Model Checking for Computation Tree Logic

In this section, the model checking algorithm for computation tree logic (CTL) is introduced. The logic CTL, which is related to branching-time behaviour, is defined in terms of its syntax and semantics. Model checking for branching time is based on a fixpoint characterisation for this logic. We recall some details about the underlying theory of fixpoints. Based on this theory, we give the characterisation of CTL formulas and depict the notion for the model checking algorithm using it.

Symbolic model checking is an approach for reducing the necessary effort and thus extend the limits of the explicit state space exploration. We show how to adapt the algorithms for this more effective symbolic, i.e., implicit representation of states. Conditional fairness, although necessary for proving liveness for distributed systems, cannot be expressed within CTL logic (see Section 3.5.2). Thus, it is a problem for CTL model checking to deal with fairness. This is addressed in the last subsection. The results surveyed in this section are taken from the literature. We refer to [CGL94], [CGP00], and [Kro98].

3.3.1   The Logic CTL

Syntax. We distinguish *state formulas*, expressing a property of a specific state, and *path formulas*, modelling a proposition over a specific path. The set $\mathcal{V}$ denotes the set of atomic propositions (i.e., boolean variables for characterising states). The syntax of a CTL formula is given by the following rules.

Definition 3.15:   Syntax of CTL

> *state formulas*:
>
> > i.) If $\varphi \in \mathcal{V}$, then φ is a state formula.
> > ii.) If φ and ψ are state formulas, then $\neg\varphi$ and $\varphi \vee \psi$ are state formulas.
> > iii.) If φ is a path formula, then $\mathbf{E}\,\varphi$ is a state formula.
>
> *path formulas*:
>
> > i.) If φ and ψ are state formulas,
> > then $\mathbf{X}\,\varphi$ and $\varphi\,\mathbf{U}\,\psi$ are path formulas.

Roughly spoken, $\mathbf{E}$ is an existential quantifier for paths, $\mathbf{X}$ refers to the next state, and $\mathbf{U}$ is an until operator for paths: $\varphi\,\mathbf{U}\,\psi$ states that φ is true until ψ gets true which includes that this happens eventually. Some additional operators are used as abbreviations:

$$
\begin{aligned}
\varphi \wedge \psi &:\Leftrightarrow \neg(\neg\varphi \vee \neg\psi) & &\text{the boolean operator } and, \\
\mathbf{F}\,\varphi &:\Leftrightarrow (true\,\mathbf{U}\,\varphi) & &\text{the temporal operator for } future, \\
\mathbf{G}\,\varphi &:\Leftrightarrow \neg\mathbf{F}\,\neg\varphi & &\text{the temporal operator for } always, \\
\mathbf{A}\,\varphi &:\Leftrightarrow \neg\mathbf{E}\,\neg\varphi & &\text{the temporal path operator } for\ all\ paths.
\end{aligned} \tag{3.2}
$$

Since CTL expresses properties over a computation tree of branching time, we are dealing with state formulas comprising the whole tree with all its possible paths rather than dealing with path formulas reflecting properties over single paths (as we do with LTL formulas). In CTL a state formula is not also a path formula (in contrast to CTL* which comprises CTL and LTL, see Section 3.5). Thus, we cannot combine the temporal operators arbitrarily: $\mathbf{E}$ and $\mathbf{A}$ are only allowed for prefixing path formulas, whereas $\mathbf{X}$, $\mathbf{U}$, $\mathbf{F}$ and $\mathbf{G}$ are only allowed in combination with state formulas. That is, the given definition of CTL syntax restricts the combination of temporal operators such that the temporal operators $\mathbf{X}$, $\mathbf{U}$, $\mathbf{F}$ and $\mathbf{G}$ must be prefixed with one of the path operators **E** or **A**. Thus, formulas of the form $(\mathbf{X}\,\varphi)$ or $(\mathbf{A}\,(\mathbf{X}\,\varphi\,\vee\,\mathbf{F}\,\psi))$ are not allowed.

We get eight basic temporal operators that can be used in CTL: **AX**, **EX**, **AG**, **EG**, **AF**, **EF**, **AU**, and **EU**. These are expressible with **EX**, **EU**, and **EG** only:

$$
\begin{aligned}
\mathbf{EF}\,\varphi &\iff \mathbf{E}\,(true\,\mathbf{U}\,\varphi) \\
\mathbf{AX}\,\varphi &\iff \neg\mathbf{EX}\,(\neg\varphi) \\
\mathbf{AG}\,\varphi &\iff \neg\mathbf{EF}\,(\neg\varphi) \iff \neg(\mathbf{E}\,(true\,\mathbf{U}\,\neg\varphi)) \\
\mathbf{AF}\,\varphi &\iff \neg\mathbf{EG}\,(\neg\varphi) \\
\mathbf{A}\,(\varphi\mathbf{U}\,\psi) &\iff \neg\mathbf{E}\,(\neg\psi\mathbf{U}\,\neg\varphi\wedge\neg\psi)\wedge\neg\mathbf{EG}\,(\neg\psi)
\end{aligned}
\tag{3.3}
$$

In the following we define the semantics for CTL formulas. In fact, only **EX**, **EG**, and **EU** are sufficient as base operators if we follow (3.3).

Semantics. The semantics of a CTL formula is related to a temporal structure. We specify a property of the corresponding computation tree by means of state formulas that are related to the initial state. (Path formulas occur only as sub-formulas in this context.) If φ is a state formula then we say that φ holds in a state s of a temporal structure M, more formally: $M, s \models \varphi$. (The structure M might be omitted if it is clear in a given context.)

Let φ, φ_1 and φ_2 be state formulas, and ψ be a path formula. (Recall that $L(s)$ is the set of labels of state s, and $\mathcal{V}$ is the set of atomic propositions.) Then we define the satisfiability relation $\models$ in the following way:

Definition 3.16:   Semantics of CTL

$$
\begin{aligned}
s \models \varphi \quad &: \Leftrightarrow \varphi \in L(s), \text{if } \varphi \in \mathcal{V} \\
s \models \neg\varphi \quad &: \Leftrightarrow s \not\models \varphi \\
s \models \varphi_1 \vee \varphi_2 \quad &: \Leftrightarrow s \models \varphi_1 \text{ or } s \models \varphi_2
\end{aligned}
$$

$$
s \models \mathbf{EX}\,\varphi \quad :\Leftrightarrow
\begin{cases}
\text{there is a path } \underline{s}\text{ , starting at state } s, \\
\text{such that } s_1 \text{ is the next state in } \underline{s} \text{ and} \\
s_1 \models \varphi \text{ holds}
\end{cases}
$$

$$
s \models \mathbf{E}\,(\varphi_1\mathbf{U}\,\varphi_2) : \Leftrightarrow
\begin{cases}
\text{there is a path } \underline{s}\text{ , starting at state } s, \\
\text{such that there exists a } k \geq 0, \text{ with } s_k \models \varphi_2, \\
\text{and for all } 0 \leq j < k,\ s_j \models \varphi_1 \text{ holds}
\end{cases}
$$

$$
s \models \mathbf{EG}\,\varphi \quad :\Leftrightarrow
\begin{cases}
\text{there is a path } \underline{s}\text{ , starting at state } s, \\
\text{such that for all } k \geq 0,\ s_k \models \varphi \text{ holds.}
\end{cases}
$$

We call a CTL state formula φ *valid* in a given temporal structure M with initial state s_0 if $M, s_0 \models \varphi$. We call φ *satisfiable* if there exists a structure M with initial state s_0 such that $M, s_0 \models \varphi$. Each M that satisfies φ is called a *model* of φ.

Example 3.1:

- **Safety:** In each state, which can be reached, *system_crash* never holds (*the bad thing will never occur*).
    $$\mathbf{AG}\,(\neg system\_crash)$$

- **Liveness:** All computation paths satisfy that if a request *req* is given, then an acknowledgement *ack* will eventually occur
    $$\mathbf{AG}\,(req \to \mathbf{AF}\,ack)$$
    or in each computation path a process is active infinitely often[3].
    $$\mathbf{AG}\,(\mathbf{AF}\,process\_active)$$

---

[3]In fact, this property can be seen as *unconditional* fairness which can be described in terms of CTL operators. However, in many applications the fairness property is forced with respect to some condition that has to be satisfied as well (see Section 3.5.2).

- **Absence of Livelock:** In every state there is a path on which
  eventually the system is able to proceed. For instance, proceeding
  may be specified as being able to read, i.e., *readable* is satisfied.
  AG (**EF** *readable*)

- **Inconsistency:** Always (on every path in every state) *False* is
  satisfied, i.e., there is no valid path in the given temporal structure.
  AG *False*

3.3.2   Model Checking CTL Formulas

For any CTL formula it is possible to give a fixpoint characterisation. This is due
to the fact that CTL formulas are satisfied by a set of states (rather than paths).
Therefore the domain is $\mathcal{P}(\mathcal{S})$, which is a complete lattice. Thus, fixpoint theory
can be exploited for the model checking problem.

A simple recursive algorithm, the fixpoint computation, can be used to determine
all states of a given structure that are reachable by applying iteratively the transition
relation and that also satisfy the given CTL formula. The computation starts with
the set of states that satisfies the (innermost) atomic sub-formula which contains no
state and paths quantifiers. Once the set of satisfying reachable states is computed
it remains to show that the initial state of the structure s_0 is an element of this set.
If this is true, then the structure is a model of the CTL formula.

We start with some preliminaries of the underlying theory and the description of
the fixpoint characterisation of CTL formulas, and continue with the corresponding
algorithms for searching the state space. In order to improve efficiency of the model
checking procedure, symbolic model checking can be applied. This is a technique
that uses a particular data structure, known as binary decision diagrams, in order to exploit their succint representation and the efficiency of the corresponding
algorithms.

Fixpoint Characterisation

Every temporal operator in CTL can be described as a fixpoint equation. This
characterisation yields the basis for the model checking algorithm used to check
CTL formulas.

Preliminaries.   Generally, a *fixpoint* of a function $f : \mathcal{A} \to \mathcal{A}$ is an element $x \in \mathcal{A}$
that satisfies $f(x) = x$. In case $\mathcal{A}$ is a lattice and f is a monotonic function, then f
has a definite least fixpoint fp_{min} as well as a definite greatest fixpoint fp_{max}[4]. If $\mathcal{A}$
is a complete lattice (i.e., $\mathcal{A}$ has a lower and an upper bound) and f is continuous
then both extremal fixpoints can be characterised as the limit of applying f to the
supremum (the least upper bound) of the lattice, or its infimum (the greatest lower
bound). If we start with the supremum, the result will be the greatest fixpoint,
starting with the infimum we get the least fixpoint. In fact, these theoretical results
yield an algorithm for computing the fixpoints. Dealing with a finite lattice with
$| \mathcal{A} | = n$, it can be shown that the procedure will terminate in at least n steps.

Given the lattice of power sets over S, namely $\mathcal{P}(\mathcal{S})$ with the ordering relation
$\subseteq$ we get the entire set S as its supremum, and $\emptyset$ as its infimum. That is, the lattice
is complete and we can order the elements as $\emptyset \subseteq S_1 \subseteq S_2 \ldots \subseteq S$ with $S_i \in \mathcal{P}(\mathcal{S})$
for all i. We call a function $f : \mathcal{P}(\mathcal{S}) \to \mathcal{P}(\mathcal{S})$

- *monotonic* if $S_i \subseteq S_j$ implies $f(S_i) \subseteq f(S_j)$

---

[4]We denote the fixpoints as functions with two arguments $fp_{min}(x, f(x))$ and $fp_{max}(x, f(x))$,
respectively, to indicate that they satisfy the equation $x = f(x)$.

- $\cup$-*continuous* if $S_1 \subseteq S_2 \ldots$ implies $f(\bigcup_i S_i) = \bigcup_i (f(S_i))$
- $\cap$-*continuous* if $S_1 \supseteq S_2 \ldots$ implies $f(\bigcap_i S_i) = \bigcap_i (f(S_i))$

Within a complete lattice, the monotonicity of a function f is equivalent to f being $\cup$-*continuous* and is also equivalent to f being $\cap$-*continuous* (see also [CGL94]).

Fixpoint Computation. If $\mathcal{P}(\mathcal{S})$ is a complete lattice and $f : \mathcal{P}(\mathcal{S}) \to \mathcal{P}(\mathcal{S})$ is $\cup$-*continuous* (and therefore monotonic as well) then f always has a least fixpoint that is computed by applying f iteratively on the infimum $\emptyset$, i.e.,

$$fp_{min}(Z, f(Z)) = \bigcup_i (f^i(\emptyset)) \qquad (3.4)$$

If f is $\cap$-*continuous* it always has a greatest fixpoint that is computed by applying f iteratively on the supremum S, i.e.,

$$fp_{max}(Z, f(Z)) = \bigcap_i (f^i(S)) \qquad (3.5)$$

These equations yield the basis for algorithmic determination. Two simple functions `least_fixpoint` and `greatest_fixpoint` compute the extremal fixpoints:

```
funct least_fixpoint f : P(S) → P(S) : P(S);
  begin
    B := ∅;
    B' := f(B);
    while B ≠ B' do
          begin
            B := B';
            B' := f(B);
          end;
    return(B);
  end.
```

The invariant of the while-loop in the body of the function is given by the assertion

$$[\mathcal{B}' = f(\mathcal{B})] \wedge [\mathcal{B}' \subseteq fp_{min}(\mathcal{B}, f(\mathcal{B}))].$$

Obviously, at the beginning of the i-th iteration of the loop, $\mathcal{B} = f^{i-1}(\emptyset)$, and $\mathcal{B}' = f^i(\emptyset)$. We get a sequence of ordered intermediate results $\emptyset \subseteq f(\emptyset) \subseteq f^2(\emptyset) \subseteq \ldots$ The maximum number of iterations before the while-loop terminates is bounded by the number of elements in the set S. When the loop terminates we get as a result $\mathcal{B} = f(\mathcal{B})$. In combination with the invariant of the loop, we conclude that $\mathcal{B} = fp_{min}(\mathcal{B}, f(\mathcal{B}))$, i.e., $\mathcal{B}$ is the least fixpoint of function f. Analogously, we can argue for the function `greatest_fixpoint`.

```
funct greatest_fixpoint f : P(S) → P(S) : P(S);
  begin
    B := S;
    B' := f(B);
    while B ≠ B' do
          begin
            B := B';
            B' := f(B);
          end;
    return(B);
  end.
```

Fixpoint Characterisation Given a temporal structure $M = (S, S_0, R, L)$, we identify every set of states S_i of M with the predicates that are true in S_i. Thus, we get a lattice $\mathcal{P}(\mathcal{S})$ of predicates over S with the set inclusion $\subseteq$ as its ordering relation. $\mathcal{P}(\mathcal{S})$ is complete, the least element is $\emptyset$ and the greatest element is S. The basis of fixpoint characterisation is a function that maps sets of states onto set of states, that is $f : \mathcal{P}(S) \rightarrow \mathcal{P}(S)$. We call f a *predicate transformer*.

For the fixpoint characterisation, we identify a CTL formula φ with the set of those states, in which φ is satisfied, i.e., $\{s \mid M, s \models \varphi\}$. Thus, we may apply a temporal logic operator not only to a temporal formula but to a set as well. This set, when being identified by a predicate, is an element of $\mathcal{P}(\mathcal{S})$. We may describe each CTL operator in terms of a least or greatest fixpoint of an appropriate predicate transformer over $\mathcal{P}(\mathcal{S})$ as follows:

$$
\begin{aligned}
\mathbf{A}\,(\varphi\,\mathbf{U}\,\psi) &= \mathit{fp}_{min}(Z, \psi \vee (\varphi \wedge \mathbf{AX}\,Z)) \\
\mathbf{E}\,(\varphi\,\mathbf{U}\,\psi) &= \mathit{fp}_{min}(Z, \psi \vee (\varphi \wedge \mathbf{EX}\,Z)) \\
\mathbf{AF}\,\varphi &= \mathit{fp}_{min}(Z, \varphi \vee \mathbf{AX}\,Z) \\
\mathbf{EF}\,\varphi &= \mathit{fp}_{min}(Z, \varphi \vee \mathbf{EX}\,Z) \\
\mathbf{AG}\,\varphi &= \mathit{fp}_{max}(Z, \varphi \wedge \mathbf{AX}\,Z) \\
\mathbf{EG}\,\varphi &= \mathit{fp}_{max}(Z, \varphi \wedge \mathbf{EX}\,Z)
\end{aligned}
\tag{3.6}
$$

The first parameter Z of fp_{min} and fp_{max}, respectively, denotes a set of states and the second parameters provide the suitable predicate transformer f applied to Z, for the corresponding CTL formula. These predicate transformers map the set Z to the union of the current states satisfying the involved predicates (given by ψ and φ) and the satisfying states reachable in the next step (as indicated by the temporal operators $\mathbf{AX}$ and $\mathbf{EX}$). This coincides with the notion of traversing the transition system and collecting all states along the paths such that the CTL formula is satisfied. When a fixpoint is found the search for all reachable states is completed.

In order to prove these fixpoint equations, we show that the predicate transformers are monotonic, and the given CTL formulas are exactly the minimal or maximal fixpoints for the corresponding transformer. In the appendix, we give the proof for $\mathbf{EG}\,\varphi$ and $\mathbf{A}\,(\varphi\mathbf{U}\,\psi)$ as examples (see appendix A.1). All other equations can be proved analogously.

In order to show how the given fixpoint characterisations can be used for determining all states satisfying a certain CTL formula, we depict the following example.

> **Example 3.2:** Given a temporal structure M (denoted as a graph in figure 3.1) we want to check whether CTL formula $\phi = \mathbf{E}\,(p\,\mathbf{U}\,q)$ is satisfied, i.e., $M, s_0 \models \mathbf{E}\,(p\,\mathbf{U}\,q)$. $\mathbf{E}\,(p\,\mathbf{U}\,q)$ is described as the least fixpoint of the predicate transformer $f(Z) := q \vee (p \wedge \mathbf{EX}\,Z)$. Using the algorithm `least_fixpoint`, we can compute the set of states in M satisfying ϕ starting the procedure with $\emptyset$. In the first iteration, we get $f(\emptyset) = q \vee (p \wedge \mathbf{EX}\,\emptyset) = q$. This equation is satisfied by state s_2 only (in figure 3.1 the states that yield a solution to the equation are filled). In the second iteration, we have to solve $f^2(\emptyset) = f(\{s_2\}) = q \vee (p \wedge \mathbf{EX}\,(\{s_2\}))$ which obviously equals to the set $\{s_2, s_1\}$. In the third step, we find that $f^3(\emptyset) = \{s_2, s_1, s_0\}$ which is the fixpoint since $f^4(\emptyset) = f^3(\emptyset)$ yields the last iteration of the loop.

Model Checking Algorithm

If a temporal structure M and a requirement specification in terms of a CTL formula φ is given then model checking involves finding all states in the set of (reachable)

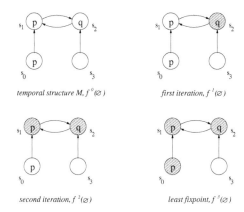

temporal structure M, f⁰(∅) — *first iteration, f¹(∅)*

second iteration, f²(∅) — *least fixpoint, f³(∅)*

Figure 3.1: Computing the set of states satisfying $\mathbf{E}\,(p\,\mathbf{U}\,q)$

states of the structure that satisfy the requirement. The set of satisfying states is denoted as $Sat(\varphi) = \{s \mid M, s \models \varphi\}$. Formally, the model checking problem is given as

$$M, s_0 \models \varphi. \tag{3.7}$$

That is, for solving the model checking problem, we have to prove that the initial state s_0 is an element of $Sat(\varphi)$.

For simple CTL formulas, we showed above how to compute $Sat(\varphi)$ by means of fixpoint computation. For more complex formulas, we rely on their semantical definition (following the definition 3.16 of $\models$): We first compute the set $Sat(\varphi_i)$ for all sub-formulas φ_i. Then these sets are combined corresponding to the semantics of the original complex formula. Thus, it suffices to know the definition of Sat applied to the basic CTL formulas.

We define $Sat(\varphi)$ as the set of states satisfying φ more accurately as follows (recalling that $L(s)$ is the set of labels of state s, and $\mathcal{V}$ is the set of atomic propositions):

Definition 3.17:   Satisfying States

$$
\begin{aligned}
Sat(\varphi) &:= \{s \mid \varphi \in L(s)\},\ \text{if}\ \varphi \in \mathcal{V} \\
Sat(\neg\varphi) &:= \{s \mid S \setminus Sat(\varphi)\} \\
Sat(\varphi_i \vee \varphi_j) &:= Sat(\varphi_i) \cup Sat(\varphi_j) \\
Sat(\mathbf{EX}\,\varphi) &:= \{s \mid \exists s'.(s' \in Sat(\varphi)) \wedge (s, s') \in R\} \\
Sat(\mathbf{E}\,(\varphi_i\,\mathbf{U}\,\varphi_j)) &:= \mathit{fp}_{min}(Z, \varphi_i \vee (\varphi_j \wedge \mathbf{EX}\,Z)) \\
Sat(\mathbf{EG}\,\varphi) &:= \mathit{fp}_{max}(Z, \varphi_i \wedge \mathbf{EX}\,Z)
\end{aligned}
$$

Following this definition, the necessary basic operations for a model checking algorithm for CTL formulas are:

- evaluation of label $L(s)$ for the variables occurring in φ

- set theoretic operations like $\setminus$ or $\cup$

- evaluation of the transition relation R

- fixpoint computation for the temporal operators $\mathbf{EU}$ and $\mathbf{EG}$.

According to the equivalences given in (3.3), all other temporal operators can be expressed in terms of the operators occurring in definition 3.17.

This notion of a checking procedure describes ordinary model checking of CTL formulas on the basis of temporal branching-time structures. In the following section, we introduce the notion of symbolic model checking, which is a special kind of a checking procedure exploiting the efficiency of algorithms on binary decision diagrams.

3.3.3   Symbolic Model Checking

The complexity of the ordinary model checking algorithm for CTL is *linear* in size of the computation graph of the structure and *linear* in size of the CTL formula to be checked. The algorithm is shown to be quite fast in practice. However, the number of states may explode when dealing with concurrent systems. Often model checking is not feasible for this reason. In some cases, we can diminish the *state explosion problem* by using a symbolic representation. In this approach, the limiting factor for the model checking problem is not the state space anymore but rather determined by the size of the *implicit* or *symbolic representation* of the computation graph, which may be much smaller. (Unfortunately, this does not hold for every problem as, for example, shown in [Bry86]).

For a *symbolic representation*, a state of the structure is encoded as a bit vector $\vec{s} : \mathbb{B}^n$, and the transition relation as a function over pairs of bit vectors accordingly $R : (\mathbb{B}^n \times \mathbb{B}^n) \to \mathbb{B}$ (see [BCM$^+$92]). The atomic propositions of $\mathcal{V}$ give the boolean *state variables* s_i that characterise a state. Thus, all variables in $\mathcal{V}$ constitute a state vector $\vec{s}$ of length $n = |\ \mathcal{V}\ |$. As usual, we identify sets of states S_i and the corresponding characterising function $\Phi : \mathbb{B}^n \to \mathbb{B}$ such that Φ is true if and only if $s \in S_i$. The boolean functions R and Φ can be represented by means of *reduced ordered binary decision diagrams* (ROBDDs), which yields a concise canonical representation (for more details about ROBDDs see Section 4.1).

Thus, the notion of *symbolic computation* is related to the use of an implicit representation of states and transitions, as those being characterised by the corresponding evaluation of state variables in $\mathcal{V}$. In order to compute $Sat(\varphi)$ by the main steps in definition 3.17, we can use very efficient algorithms for computing boolean operations on ROBDDs. In the following the adapted model checking algorithms are introduced.

Again, the goal is to find all those states that satisfy a given CTL formula φ, that is $Sat(\varphi)$. To compute this set of states, we use a procedure **Check** applied to parameter φ. The resulting set of states is characterised by the boolean function $\Phi(\vec{s})$ (i.e., Φ is the characteristic function of $Sat(\varphi)$). We get the following equation:

$$Sat(\varphi) = \texttt{Check}(\varphi) = \Phi(\vec{s}) \qquad (3.8)$$

We compute the set inductively over the structure of the formula. Since we are dealing with characteristic functions, only boolean operators are necessary for simple formulas:

$$\begin{aligned}
\texttt{Check}(\varphi) &:= \Phi(\vec{s}) \\
\texttt{Check}(\neg\varphi) &:= \neg\texttt{Check}(\varphi) \\
\texttt{Check}(\varphi \vee \psi) &:= \texttt{Check}(\varphi) \vee \texttt{Check}(\psi)
\end{aligned} \qquad (3.9)$$

For more complex formulas we introduce special procedures, namely **CheckEX**, **CheckEU**, and **CheckEG**. Following the equivalences given in (3.3) these are sufficient for treating every CTL operator.

- For the next operator $\mathbf{EX}(\varphi)$ we have to find all those preceding states for which one of the successor states satisfies φ. With $\texttt{Check}(\mathbf{EX}\,\varphi) = \texttt{CheckEX}(\texttt{Check}(\varphi)) = \texttt{CheckEX}(\Phi(\vec{s}))$ we get[5]

$$\texttt{CheckEX}(\Phi(\vec{s})) \;:=\; (\exists \vec{s}'\,.(\Phi(\vec{s}') \wedge R(\vec{s}, \vec{s}'))) \tag{3.10}$$

Since $\Phi(\vec{s}')$ as well as $R(\vec{s}, \vec{s}')$ are represented as ROBDDs we can exploit the basic operations on OBDD structures. These are implemented as very efficient algorithms. For computing the solution of equation (3.10) we use the algorithm for *relational product*. This is explained in detail in Section 4.1.1 on page 53.

The procedures $\texttt{CheckEU}$, and $\texttt{CheckEG}$ use additionally the algorithms for fixpoint computation as introduced with the functions $\texttt{least\_fixpoint}$ and $\texttt{greatest\_fixpoint}$. Within both we have to search the entire state space.

- With the fixpoint characterisation in (3.6) and $\texttt{Check}(\mathbf{E}\,(\varphi_1\,\mathbf{U}\,\varphi_2)) = \texttt{CheckEU}(\texttt{Check}(\varphi_1), \texttt{Check}(\varphi_2)) = \texttt{CheckEU}(\Phi_1(\vec{s}), \Phi_2(\vec{s}))$ we get

$$\begin{aligned} \texttt{CheckEU}(\Phi_1(\vec{s}), \Phi_2(\vec{s})) := \\ \mathit{fp}_{min}(Z(\vec{s}), \Phi_2(\vec{s}) \vee (\Phi_1(\vec{s}) \wedge \texttt{CheckEX}(Z(\vec{s})))) \end{aligned} \tag{3.11}$$

Again, the parameters $\Phi_1(\vec{s})$ and $\Phi_2(\vec{s})$ are characteristic functions (i.e., boolean functions mapping $\mathbb{B}^n \to \mathbb{B}$) that are internally represented as ROBDDs. The algorithm for computing the set $\texttt{CheckEU}(\Phi_1(\vec{s}), \Phi_2(\vec{s}))$ is simply a while loop, using only boolean operations on ROBDDs and the function for computing $\texttt{CheckEX}(\Phi(\vec{s}))$.

---

funct CheckEU $(\Phi_1(\vec{s}), \Phi_2(\vec{s})) : \mathbb{B}^n \to \mathbb{B}$;
  begin
    $B(\vec{s}) := \textsf{False}$;
    $B'(\vec{s}) := \Phi_2(\vec{s})$;
    while $B(\vec{s}) \neq B'(\vec{s})$ **do**
      begin
        $B(\vec{s}) := B'(\vec{s})$;
        $B'(\vec{s}) := \Phi_2(\vec{s}) \vee (\Phi_1(\vec{s}) \wedge \texttt{CheckEX}(B(\vec{s})))$;
      end;
    return$(B(\vec{s}))$;
  end.

The intermediate set $B(\vec{s})$ is initially $\textsf{False}$, which is the characteristic function for $\emptyset$. $B'(\vec{s})$ is initially set to $\Phi_2(\vec{s})$ which equals fp_{min} applied to $\textsf{False}$ in equation (3.11).

In the j-th iteration of the loop, B^j determines the set of states in which a state s_j is reachable in at most j steps such that $\Phi_2(\vec{s}_j)$ is true (i.e., φ_2 holds in state s_j) and for all states s_i with $i < j$ (i.e., for all states before in the path) $\Phi_1(\vec{s}_i)$ is true (i.e., φ_1 holds in states s_i). This is exactly the set of states satisfying *"there is a path such that φ_1 holds until φ_2 is true"* and matches the semantics of the formula $\mathbf{E}\,(\varphi_1\,\mathbf{U}\,\varphi_2)$ to be checked.

---

[5]Usually Φ is not given in terms of primed variables. Therefore, in the literature $\Phi(\vec{s}')$ is denoted as $\Phi(\vec{s})|_{\vec{s}:=\vec{s}'}$ which represents the term that results from renaming all occurrences of unprimed variables in $\Phi(\vec{s})$ by their primed counterparts. We choose the different notation here in order to ease the understanding.

- For the CTL operator **EG** , we get according to the fixpoint characterisation in (3.6), and with $\mathtt{Check}(\mathbf{EG}\,(\varphi)) = \mathtt{CheckEG}(\mathtt{Check}(\varphi)) = \mathtt{CheckEG}(\Phi(\vec{s}))$, the following equation

$$\mathtt{CheckEG}(\Phi(\vec{s})) := \mathit{fp}_{max}(Z(\vec{s}), \Phi(\vec{s}) \wedge \mathtt{CheckEX}(Z(\vec{s}))) \tag{3.12}$$

With the parameter $\Phi(\vec{s})$ represented as an ROBDD (as well as the intermediate results $B(\vec{s})$), and the start value **True** representing the entire set of states, we adapt the algorithm **greatest_fixpoint** for computing the set of all satisfying states.

funct CheckEG $(\Phi(\vec{s})) : \mathbb{B}^n \to \mathbb{B}$;
  begin
    $B(\vec{s}) := \mathtt{True}$;
    $B'(\vec{s}) := \Phi(\vec{s})$;
    while $B(\vec{s}) \neq B'(\vec{s})$**do**
      begin
        $B(\vec{s}) := B'(\vec{s})$;
        $B'(\vec{s}) := \Phi(\vec{s}) \wedge \mathtt{CheckEx}(B(\vec{s}))$;
      end;
    return$(B(\vec{s}))$;
  end.

After the i-th iteration of the loop the function $B(\vec{s})$ characterises the set of states satisfying φ in all states that are reachable in i or less steps. The algorithm terminates if with each next step only those states are reachable that are already part of the set, i.e., a fixpoint is reached.

3.3.4 Fairness Constraints

In concurrent systems, often the requirements are provable only under the assumption of *fairness*. For example, the **mutex**-problem is solvable only if we are assuming that the scheduling of the concurrent processes is fair in that none of the processes is suspended forever (for a proof see [KW97]). That is, in order to prove liveness, we have to restrict our attention to the fair behaviour of the scheduler.

Generally, fairness means that some properties, called *fairness constraints*, hold infinitely often along all paths. Each path that satisfies these conditions is called a fair path. (For a more elaborated introduction see 3.5.2). Since fairness itself is not expressible directly in CTL logic (see also 3.5.2) but only the fairness constraints, we have to extend the algorithms for CTL model checking. The goal is to filter the set of all possible computations in order to exclude all unfair paths.

Let $\Xi = \{\xi_1, \ldots, \xi_n\}$ be a set of fairness constraints that are given as CTL formulas. As the set of states satisfying any constraint ξ_i is characterised by $Sat(\xi_i)$, the set of fair states satisfying the requirement φ is given by $Sat(\varphi, \xi_1, \ldots, \xi_n)$. The extension of the checking algorithm is based on an extended fixpoint characterisation for **EG** (φ), **EX** (φ), and **E** $(\varphi \, \mathbf{U} \, \psi)$. The following new procedures **CheckEGFair**, **CheckEXFair**, and **CheckEUFair** are required.

1. **CheckEGFair**:
   All states in $Sat(\mathbf{EG}\,\varphi, \Xi)$ have to satisfy two conditions:
   - all states in $Sat(\mathbf{EG}\,\varphi, \Xi)$ have to satisfy φ,
     i.e., $Sat(\mathbf{EG}\,\varphi, \Xi) \subseteq Sat(\mathbf{EG}\,\varphi)$

– for all fairness constraints ξ_i and all states $s \in Sat(\mathbf{EG}\,\varphi, \Xi)$ there is a
path of length $n \geq 1$ from s to some state in $Sat(\mathbf{EG}\,\varphi, \Xi)$ satisfying ξ_i
such that all other states on the path satisfy φ, i.e., they are elements of
$Sat(\varphi)$.

As a consequence of these conditions, we extend equation (3.12) to the following fixpoint equation for `CheckEGFair`:

$$\begin{aligned} &\texttt{CheckEGFair}(\Phi(\vec{s}),\,\Xi) := \\ &\quad \textit{fp}_{max}\left[\, Z, \Phi(\vec{s}) \wedge \bigwedge_{i=1}^{n} \texttt{CheckEX}\big(\texttt{CheckEU}(\Phi(\vec{s}), Z(\vec{s}) \wedge \xi_i(\vec{s}))\big)\right] \end{aligned} \qquad (3.13)$$

The fixpoint evaluation can be done in the same way as before except that
several nested fixpoint computations are necessary (inside the computation of
the until operator, see function `CheckEU`, 3.11 on page 35).

Checking $\mathbf{EX}\,\varphi$ and $\mathbf{E}\,(\varphi\,\mathbf{U}\,\psi)$ under fairness constraints is simpler if we compute
first of all the set of states of structure M which are the initial states for the fair
paths. This can be done with the equation for `CheckEGFair` introduced above.

$$S_{fair}(\Xi) = \texttt{CheckEGFair}(\texttt{True}, \Xi) \qquad (3.14)$$

2. **CheckEXFair**:
The formula $\mathbf{EX}\,\varphi$ under fairness constraints Ξ is true in state s if and only if
there is a successor state s' with $s' \models \varphi$ and s' being the start of a fair path.
If the set of initial states of fair paths $S_{fair}(\Xi)$ is already computed (given in
terms of the characteristic function), then we can reduce the algorithm to an
equation without fairness constraints:

$$\texttt{CheckEXFair}(\Phi(\vec{s}),\,\Xi) = \texttt{CheckEX}(\Phi(\vec{s}) \wedge S_{fair}(\Xi)) \qquad (3.15)$$

3. **CheckEUFair**:
A similar argument holds for checking a formula $\mathbf{E}\,(\varphi\,\mathbf{U}\,\psi)$ under fairness
constraints. $\Psi(\vec{s})$ denotes the characteristic function for the set of states
satisfying ψ. By exploiting the set $S_{fair}(\Xi)$ we can avoid dealing with the
fairness conditions ξ_i when processing the checking algorithm. We get

$$\texttt{CheckEUFair}(\Phi(\vec{s}), \Psi(\vec{s}), \Xi) = \texttt{CheckEU}(\Phi(\vec{s}), \Psi(\vec{s}) \wedge S_{fair}(\Xi)) \qquad (3.16)$$

This completes the basic theory of model checking for CTL which we want to use
for analysing ASM models. However, our experiments, when treating case studies
(as given in Chapter 7), show the limitations of CTL model checking. We conclude
that the use of the linear time temporal logic LTL would nicely complement the CTL
approach. Therefore, we describe the theoretical background of model checking for
LTL in the next section. We focus here on one particular approach given in the
literature that allows the integration of LTL model checking into the framework of
CTL checking.

3.4   Model Checking for Linear-Temporal Logic

In this section, we consider the model checking problem for linear-temporal logic (LTL). One basic approach for LTL model checking is the construction of a *tableau* for the formula to be checked.

A tableau is a temporal structure that represents all models of the formula. Thus, tableau construction provides a means for testing satisfiability of the formula. For model checking a system M against its (temporal logic) requirements φ, we build the composition of M and the tableau of the formula T_φ. The composition of M and T_φ contains all paths of the structure that are also paths in the tableau (see the early work from Lichtenstein and Pnueli [LP85]).

Another approach for model checking LTL formulas was introduce by Vardi and Wolper et.al.([VW86], [GPVW95]). Instead of constructing a tableau for the LTL formula, their basic idea was to build an ω-automaton that accepts all models of the formula. If the system description is given as an automaton as well, then model checking can be done by checking language containment of both automaton (see also the introduction in section 3.2.2).

Both approaches suffer from the exponential complexity caused by the construction of the tableau and the ω-automaton, respectively. The construction is exponential in the size of the formula. Since in many cases the formula for expressing the common requirements is not very long, these approaches may provide good results as well but still, in the competition between CTL and LTL approaches, this is the major drawback. (The comparison is discussed in more detail in section 3.5.)

Burch et.al.([BCM$^+$92]) developed a *symbolic* algorithm for LTL model checking. By slightly modifying the original tableau construction, a symbolic representation with BDDs is rendered applicable. The approach attains additional reduction in space and time for the construction and in some cases even for the size of the tableau itself. This is also exploited in [CGH94] and [CGH97] in order to combine both CTL and LTL model checking within one framework. The core checking algorithm is the same as for CTL (as introduced in section 3.3).

In this section, we introduce the linear-temporal logic, its symbolic tableau construction and the symbolic representation of tableaus. The outcome of this procedure can be treated by a fixpoint-based model checking algorithm. It benefits from the more efficient representation, and aims at a combination of both logics being used within one checking tool. We follow the description given in [CGL94], [CGH97], and [CGP00].

3.4.1   The Logic LTL

Linear-temporal logic (LTL) is related to linear rather than branching time. Thus, the semantics of the temporal operators is given in terms of a path rather than a computation tree. Each state in a path has a unique successor state, i.e., the transition relation of the corresponding temporal structure is a function. We recall here only the basic operators for LTL, for more details see [Eme90].

Syntax.   With $\mathcal{V}$ as the set of atomic propositions (i.e., boolean variables for characterising states in a structure) the syntax of an LTL formula is given by the following rules.

Definition 3.18:   Syntax of LTL

     i.) If $\varphi \in \mathcal{V}$, then φ is a formula.

     ii.) If φ and ψ are formulas, then $\neg\varphi$ and $\varphi \vee \psi$ are formulas.

     iii.) If φ and ψ are formulas, then $\varphi \mathbf{U} \psi$ and $\mathbf{X} \varphi$ are formulas.

Given these basic operators, we can create the usual operators of propositional logic as well as the temporal operators $\mathbf{F}$ and $\mathbf{G}$:

$$\begin{aligned} \mathbf{F}\,\varphi &:\Leftrightarrow \mathbf{true}\,\mathbf{U}\,\varphi \\ \mathbf{G}\,\varphi &:\Leftrightarrow \neg\mathbf{F}\,\neg\varphi \end{aligned} \qquad (3.17)$$

In LTL logic, we do not distinguish between state and path formulas as in CTL logic since all formulas are related to paths. This becomes more clear in the next paragraph when defining the semantics.

Semantics. The semantics of LTL logic is related to a linear-temporal structure. Each computation is given as a path with each state having a determined successor. The specified behaviour may comprise a *set* of those paths instead of being a computation *tree* as in branching-time logic.

Given a *linear* structure M as in definition 3.14, and a path $\underline{s} := s_0, s_1, s_2, \ldots$ in M, we define the relation $\models$ in the following way. Let φ, φ_1 and φ_2 be path formulas, and L the labelling function of M.

Definition 3.19: Semantics of LTL

$$\begin{aligned} \underline{s} &\models \varphi & :&\Leftrightarrow s \text{ is the first state of } \underline{s} \text{ and } \varphi \in L(s), \text{ if } \varphi \in \mathcal{V} \\ \underline{s} &\models \neg\varphi & :&\Leftrightarrow \underline{s} \not\models \varphi \\ \underline{s} &\models \varphi_1 \vee \varphi_2 & :&\Leftrightarrow \underline{s} \models \varphi_1 \text{ or } \underline{s} \models \varphi_2 \\ \underline{s} &\models \mathbf{X}\,\varphi & :&\Leftrightarrow \underline{s}^1 \models \varphi \\ \underline{s} &\models \varphi_1\,\mathbf{U}\,\varphi_2 & :&\Leftrightarrow \begin{cases} \text{there exists a } k \geq 0 \text{ such that } \underline{s}^k \models \varphi_2 \\ \text{and for all } 0 \leq j < k,\ \underline{s}^j \models \varphi_1 \end{cases} \end{aligned}$$

We call an LTL formula φ *valid* in a given linear structure M with the path $\underline{s}$ if $M, \underline{s} \models \varphi$. We call φ *satisfiable* if there exists a linear structure M with path $\underline{s}$ such that $M, \underline{s} \models \varphi$. Each M that satisfies φ is called a *model* of φ.

3.4.2 Model Checking LTL Formulas

We now consider the model checking problem for LTL. In order to check systems that are given as temporal structures over branching time, we conceive each LTL formula φ as quantified over all paths, i.e., we are treating formulas of the form $\mathbf{A}\,\varphi$. In general, given a structure M the model checking problem is to find all states $s \in S$ such that $M, s \models \mathbf{A}\,\varphi$. According to the definition of $\mathbf{A}$ this is equivalent to $M, s \models \neg\mathbf{E}\,\neg\varphi$. That is, it is sufficient to be able to check formulas of the form $\mathbf{E}\,\varphi'$. If we can find a witness for $\neg\varphi$, then obviously $\mathbf{A}\,\varphi$ is not satisfied.

Since LTL formulas are paths formulas, the state-based approach of CTL checking cannot be used here. Instead we construct a special temporal structure T, called the *tableau* of φ. T includes every path that satisfies φ. By composing M and T in a suitable way, we get a structure P comprising all paths that appear in both structures. A state in M will satisfy $\mathbf{E}\,\varphi$ if and only if it is the start of a path in the composition structure P. That is, in order to check $\mathbf{E}\,\varphi$ we can simply apply the CTL model checking procedure to the composed structure P.

Tableau Construction

Let $\mathcal{V}_\varphi$ be the set of propositions occurring in φ. The tableau associated with the formula φ is a structure $T = (S_T, R_T, L_T)$, where S_T is the set of states, R_T the transition relation, and L_T the labelling function. Each state in a tableau T

is labelled with sets of *elementary formulas* $el(\varphi)$. These are obtained from the sub-formulas of φ according to the following recursive definition.

$$
\begin{aligned}
el(\varphi_1) &= \{\varphi_1\} \text{ if } \varphi_1 \in \mathcal{V}_\varphi \\
el(\neg\varphi_1) &= el(\varphi_1) \\
el(\varphi_1 \vee \varphi_2) &= el(\varphi_1) \cup el(\varphi_2) \\
el(\mathbf{X}\,\varphi_1) &= \mathbf{X}\,\varphi_1 \cup el(\varphi_1) \\
el(\varphi_1 \mathbf{U}\,\varphi_2) &= \mathbf{X}\,(\varphi_1 \mathbf{U}\,\varphi_2) \cup el(\varphi_1) \cup el(\varphi_2)
\end{aligned}
\tag{3.18}
$$

That is, besides the atomic propositions we get the next-formulas $\mathbf{X}\,\varphi$ and $\mathbf{X}\,(\varphi_1 \mathbf{U}\,\varphi_2)$ as elementary formulas. For a symbolic representation, we simply have to introduce additional boolean variables for these kind of elementary formulas.

The set of states in the tableau S_T is given as $\mathcal{P}(el(\varphi))$. The labelling function L_T labels each state with the set of atomic propositions contained in the state, i.e., $L_T : S_T \rightarrow \mathcal{P}(\mathcal{V}_\varphi)$. Now we have to conceive the transition relation R_T such that each elementary formula of a state is satisfied in that state. For this concern the function Sat_T is defined that associates with each formula φ the set of satisfying states, $Sat_T : \varphi \rightarrow \mathcal{P}(S_T)$. This is defined recursively over the contributing elementary formula φ_i.

$$
\begin{aligned}
Sat_T(\varphi_1) &= \{s \mid \varphi_1 \in s\} \text{ where } \varphi_1 \in el(\varphi) \text{ and } s \in S_T \\
Sat_T(\neg\varphi_1) &= \{s \mid s \notin Sat_T(\varphi_1)\} \\
Sat_T(\varphi_1 \vee \varphi_2) &= Sat_T(\varphi_1) \cup Sat_T(\varphi_2) \\
Sat_T(\varphi_1 \mathbf{U}\,\varphi_2) &= Sat_T(\varphi_2) \cup \big(Sat_T(\varphi_1) \cap Sat_T(\mathbf{X}\,(\varphi_1 \mathbf{U}\,\varphi_2))\big)
\end{aligned}
\tag{3.19}
$$

Since the transition relation should preserve the truth value of the elementary formulas in each state, the occurrence of next-formulas has to be reflected. That is, if $\mathbf{X}\,\varphi_i$ is true in some state s then *all* its successors s' should satisfy φ_i. Furthermore, since we are dealing with LTL formulas (which are implicitly quantified over all paths), if $\mathbf{X}\,\varphi_i$ is not in s, then $\neg\mathbf{X}\,\varphi_i$ is in s, and *no* successor of s should satisfy φ_i. Following this intention, R_T is defined as

$$
R_T(s,s') = \bigwedge_{\mathbf{X}\,\varphi_i \in el(\varphi)} s \in Sat_T(\mathbf{X}\,\varphi_i) \Leftrightarrow s' \in Sat_T(\varphi_i)
\tag{3.20}
$$

We give an example in order to illustrate the notion of tableaus.

> **Example 3.3:** Consider the formula $\varphi = a\mathbf{U}\,b$. The elementary formulas are $el(\varphi) = \{a, b, \mathbf{X}\,(a\mathbf{U}\,b)\}$. The states of T are given as $S_T = \mathcal{P}(el(\varphi))$. We get eight states, each of them labelled with a possible subset of $el(\varphi)$ as depicted in figure 3.2. Since $el(\varphi)$ contains only one sub-formula of the kind $\mathbf{X}\,\varphi_i$ the transition relation is given as $R_T = \{(s, s') \mid s \in Sat_T(\mathbf{X}\,(a\mathbf{U}\,b)) \Leftrightarrow s' \in Sat_T(a\mathbf{U}\,b)\}$ with $Sat_T(\mathbf{X}\,(a\mathbf{U}\,b)) = \{s_1, s_2, s_3, s_5\}$ and $Sat_T(a\mathbf{U}\,b) = \{s_1, s_2, s_3, s_4, s_6\}$. This gives us the first half of the state transitions. The corresponding complements, the set $S_T \backslash Sat_T(\mathbf{X}\,(a\mathbf{U}\,b)) = \{s_4, s_6, s_7, s_8\}$ and the set $S_T \backslash Sat_T(a\mathbf{U}\,b) = \{s_5, s_7, s_8\}$ provide the remaining transitions.

The given definition of R_T (3.20) may lead to a tableau containing paths that are not a model for the formula. This can happen for two reasons:

- Not every path in T needs to satisfy φ. If it contains sub-formulas of the form $(a\mathbf{U}\,b)$ then R_T does not guarantee by definition that b will eventually hold. In figure 3.2, the path looping forever in state s_3 provides an example of this.

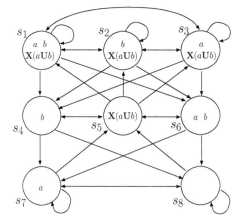

Figure 3.2: Tableau for $\varphi = a\,\mathbf{U}\,b$

- T may contain dead-end states that have no successor. For instance, there is no successor for a state labelled with $\{a, \mathbf{X}\,a, \mathbf{X}\,\neg a\}$ because a and $\neg a$ cannot be satisfied simultaneously in any next state. Since the semantics for LTL formulas is related to infinite paths, however, we may simply exclude the finite paths by removing dead-end states without loss of satisfying paths.

For the guarantee that sub-formulas of the form $a\,\mathbf{U}\,b$ are satisfied, an additional condition is necessary:

> Let T be a tableau for the formula φ, and $\underline{s} = s_0, s_1, s_2, \ldots$ a path in T with $s_0 \in Sat_T(\varphi)$. Then $\underline{s}$ will satisfy φ if and only if for every sub-formula $(a\,\mathbf{U}\,b)$ of φ and for every state s_i in $\underline{s}$ the following condition holds: if $s_i \in Sat_T(a\,\mathbf{U}\,b)$ then either $s_i \in Sat_T(b)$ or there is a later state s_j on path $\underline{s}$ (with $j > i$) such that $s_j \in Sat_T(b)$.

This condition can be re-formulated as the fairness constraint $Sat_T(\neg(a\,\mathbf{U}\,b) \vee b)$. It specifies the states that either satisfy $(a\,\mathbf{U}\,b)$ and b as well, or the negation of $(a\,\mathbf{U}\,b)$. In order to guarantee that b will eventually occur, every path has to meet these states infinitely often. This is the case for all paths that are *fair* with respect to this fairness constraint. This will be reflected later in the checking procedure.

A key property of the construction is that for every path in the temporal structure (which models the system) that satisfies φ, there is a corresponding path in the tableau of φ. To make the notion of "corresponding path" precise, a label function on paths and their restriction is given.

Let $\underline{s} = s_0, s_1, s_2, \ldots$ be a path in the temporal structure M with atomic propositions $\mathcal{V}_{\mathcal{M}}$ and label function L_M on its states. A labelling function on paths is given by $label(\underline{s}) = L_M(s_0), L_M(s_1), L_M(s_2), \ldots$ We may restrict the path labels to atomic propositions contained in φ, denoted as $\mathcal{V}_\varphi \subseteq \mathcal{V}_{\mathcal{M}}$. This is denoted as $label(\underline{s})\big|_{\mathcal{V}_\varphi} = L'_M(s_0), L'_M(s_1), L'_M(s_2), \ldots$ with $L'_M(s_i) = L_M(s_i) \cap \mathcal{V}_\varphi$ for every i. Now, the core issue of tableau construction can be stated in the following way:

Theorem 3.1. *Let T be the tableau for the LTL formula φ. Then for every temporal structure M and every path $\underline{s}$ in M, if $M, \underline{s} \models \varphi$ then there is a path $\underline{s}'$ in T that starts in a state $s_0 \in Sat_T(\varphi)$ such that $label(\underline{s})\big|_{\mathcal{V}_\varphi} = label(\underline{s}')$.*

The proof is skipped here but can be found in [CGH97].

The Product Structure

Following theorem 3.1, a product structure of M and the tableau T, which is constructed in a suitable way, will comprise the paths that are in M *and* in T. That is, the paths of M that satisfy φ.

Let $M = (S_M, R_M, L_M)$ and $T = (S_T, R_T, L_T)$ be the tableau for formula φ, which contains the atomic propositions $\mathcal{V}_\varphi$. The product structure is built according to the following definition:

Definition 3.20: Product Structure

The product structure $P = (S_P, R_P, T_P)$ consists of

- a set of states
  $S_P = \{(p, q) \mid p \in S_T, \ q \in S_M, \ \text{and} \ L_M(q) \cap \mathcal{V}_\varphi = L_T(p)\}$
- the transition relation
  $R_T = \{((p, q), (p', q')) \mid (p, p') \in R_T \ \text{and} \ (q, q') \in R_M\}$
- the labelling function
  $L_P((p, q)) = L_T(p)$

P may have dead-end states and thus finite paths, even if the tableau T contains none. On the other hand, P contains exactly the infinite paths $\underline{s}''$ for which there are infinite path $\underline{s}'$ in T and $\underline{s}$ in M such that $label(\underline{s}'') = label(\underline{s}') = label(\underline{s})\big|_{\mathcal{V}_\varphi}$. Thus, with the same argument as for the tableau, we may remove the dead-end states without losing satisfying paths since all models of φ have to be infinite. We extend the function Sat_T to Sat_P such that for every state $(p, q) \in S_P$, $(p, q) \in Sat_P(\varphi)$ holds if and only if $p \in Sat_T(\varphi)$. We give an example for a product structure.

Example 3.4: Consider the temporal structure given in figure 3.3

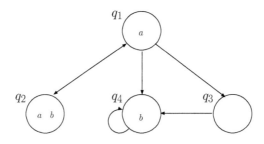

Figure 3.3: The structure M

Combining M with tableau T for $\varphi = (a \mathbf{U} b)$, as given in figure 3.2, we get the set of states $S_P = \{(s_1, q_2), (s_2, q_4), (s_3, q_1), (s_4, q_4),$

$(s_5, q_3), (s_6, q_2), (s_7, q_1), (s_8, q_3)$. The transition relation R_P is depicted in figure 3.4. For the states (s_4, q_4) and (s_8, q_3) we can see that they have no outgoing edges. These are dead-end states and can be deleted.

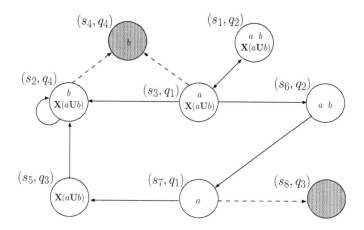

Figure 3.4: Product structure of M and tableau T

Model Checking the Product Structure

Recall the model checking problem for LTL: Given a branching-time structure M and an LTL formula of the form $\mathbf{A}\,\psi$, the model checking problem is stated as $M, s \models \mathbf{A}\,\psi$. For technical reasons, this is re-formulated according to the equivalence $(M, s \models \mathbf{A}\,\psi) \Leftrightarrow (M, s \models \neg\mathbf{E}\,\neg\psi)$. Thus, formulas of the form $\mathbf{E}\,\varphi$ with $\varphi = \neg\psi$ are investigated. If we can find a state that is the start of a path that satisfies φ, then this state and the outgoing path will provide a counterexample for $\mathbf{A}\,\psi$. If we construct a product structure P from the branching-time structure M and the tableau T_φ, then all path in P are paths in M and they also satisfy φ (see definition 3.20).

According to the discussion in the last subsection, we have to consider fairness constraints. We want to find all states in the product structure P that are the start of an infinite *fair* path with respect to the fairness constraint below.

$$\{Sat_P(\neg(a\,\mathbf{U}\,b) \vee b) \mid (a\,\mathbf{U}\,b) \text{ occurs in } \varphi\} \qquad (3.21)$$

This can be done by applying fair CTL model checking to P with the CTL formula $\mathbf{EG}\,true$ and the fairness constraints as given above. We will get all states of P that are the start of an infinite path with the property that for every sub-formula $(a\mathbf{U}\,b)$ it is guaranteed that b will be eventually true. This is summarised in theorem 3.2.

Theorem 3.2. $M, s' \models \mathbf{E}\,\varphi$ if and only if there is a state $s \in T$ such that $(s, s') \in Sat_P(\varphi)$ and $P, (s, s') \models \mathbf{EG}\,true$ under fairness constraints $\{Sat_P(\neg(a\,\mathbf{U}\,b) \vee b) \mid (a\,\mathbf{U}\,b)$ occurs in $\varphi\}$.

Again, the proof is skipped here; the reader is referred to [CGH97].

43

In our example above, it is easy to see that in the structure P (as shown in figure 3.4) the fairness constraint is fulfilled for the states $\{(s_1, q_2), (s_2, q_4), (s_3, q_1), (s_5, q_3), (s_6, q_2), (s_7, q_1)\}$. Thus, every state in P (except the removed dead-end states that are filled in the figure) satisfies $\mathbf{EG}\ true$ under the fairness constraint. However, only the states $\{(s_1, q_2), (s_2, q_4), (s_3, q_1), (s_6, q_2)\}$ are in $Sat_P(\varphi)$. Therefore, only the states $q_1, q_2,$ and q_4 in M satisfy $\mathbf{E}\,\varphi = \mathbf{E}\,(a\,\mathbf{U}\,b)$.

3.4.3   Symbolic Model Checking for LTL

To exploit a more concise symbolic representation for LTL model checking, we have to define the transition relations of M and T in terms of boolean functions which can be represented as ROBDDs. It is already described in section 3.3.3 how this can be done for the temporal structure M. In this section, we give a notion of how to represent a tableau in terms of ROBDDs.

Generally, for each elementary formula $\varphi_i \in el(\varphi)$, we introduce a boolean variable v_i. Thus, the tuple $\overline{v} = (v_1, v_2, \ldots)$ describes a state in T, and the transition relation can be given in terms of $\overline{v}$ and its copy $\overline{v}'$ (denoting truth values in the next state).

In order to get distinguishable variables, we denote with $\overline{p} = (p_1, p_2, \ldots)$ the tuple of boolean variables that represent the atomic propositions $\varphi_i \in \mathcal{V}_\varphi$. With $\overline{r} = (r_1, r_2, \ldots)$ we will denote the tuple of boolean variables representing the remaining elementary formulas $el(\varphi) \backslash \mathcal{V}_\varphi$. Within M, we will use $\overline{q}$ as a boolean vector for $\mathcal{V}_\mathcal{M} \backslash \mathcal{V}_\varphi$.

Then, a state $s \in S_T$ can be given as a pair of boolean tuples $(\overline{p}, \overline{r})$. The transition relation in the tableau R_T will be represented as a boolean formula over pairs $(\overline{p}, \overline{r})$ and $(\overline{p}', \overline{r}')$. The transition relation in the temporal structure R_M is given as a boolean function over $(\overline{p}, \overline{q})$ and $(\overline{p}', \overline{q}')$. The transition relation of the product structure R_P is the conjunct of both boolean functions:

$$R_P(\overline{p}, \overline{q}, \overline{r}, \overline{p}', \overline{q}', \overline{r}') = R_R(\overline{p}, \overline{r}, \overline{p}', \overline{r}') \wedge R_M(\overline{p}, \overline{q}, \overline{p}', \overline{q}') \tag{3.22}$$

Now the symbolic CTL model checking algorithms for treating fairness constraints (as described in section 3.3.3) can be used to find all states in P that satisfies ($\mathbf{EG}\ true$) with respect to the fairness constraint given in 3.21. The states comprise the fixpoint.

Considering the dead-end states in P, and the corresponding finite paths that are not a proper model for the LTL formula, it can be shown that these are excluded by means of the fixpoint computation under fairness constraints: For every state s that is in the fixpoint of the predicate transformer for ($\mathbf{EG}\ true$) we know that there is a path from s meeting infinitely often states s' that satisfy the fairness constraints. That is, we only get those states, which have infinite paths meeting all fairness constraints infinitely often. None of the dead-end states can fulfil this property.

The resulting states are given as boolean vectors $(\overline{p}, \overline{r}, \overline{q})$. A state $(\overline{p}, \overline{q}) \in M$ satisfies ($\mathbf{EG}\ true$) if and only if there exists a vector $\overline{r}$ such that $(\overline{p}, \overline{r}, \overline{q})$ is in the fixpoint and $(\overline{p}, \overline{r}) \in Sat_T(\varphi)$.

With this last subsection, we finish the description of a model checking approach for LTL that can be nicely integrated into symbolic model checking of CTL. This approach suggests an integration of both logics into one model checking tool. In the following section, we given a comparison of CTL and LTL that should motivate the use of a combination of both logics.

3.5   Linear-Time versus Branching-Time Logic

Comparing LTL with CTL requires two measurements:

1. The complexity of the checking procedure. This measure depends on both, the size of the formula and the size of the structure to be checked.

2. The expressiveness of the language given with the logic.

In order to motivate our decision to provide ASM with the checking facility for both kinds of temporal logic, we gather some remarks concerning complexity and expressiveness in the following subsections.

3.5.1   Complexity of the Algorithms

Model checking CTL formulas is *linear* in the size of the temporal structure M and *linear* in the size of the temporal formula φ. This renders the CTL approach feasible for quite large systems. However, if the system is of exponential size the complexity of the checking procedure is exponential, as well (e.g., a switch over a data domain with n elements has 2^n different states).

If symbolic representation is used by means of binary decision diagrams (BDDs), then the checking algorithm is operating not on the system states (or sets of states) directly, but rather on BDDs that represent the corresponding characteristic functions for sets of states. That is, the complexity depends on the size of the BDD instead. In many cases, the BDD size is much smaller than the size of the structure itself. But this does not hold for every example (see [Bry86]). The BDD structure may grow to exponential size as well, and in some cases may be even larger than the explicit representation.

The complexity of the linear-time approach is *linear* in the size of the structure but *exponential* in the size of the logical formula. This is caused by the construction of the tableau that represents the requirements (i.e., the LTL formula). This drawback of the linear-time approach, however, is not absolute.

Firstly, the algorithms can be improved. In the traditional approach the construction of the tableau is guided by the appearance of the LTL formula, i.e., by its logical operators (see [Wol81] or [BAMP81] for example). When adapting the construction principle, however, the benefit of symbolic representation can be exploited. [BCM+92], [CGH94] and [CGH97] introduce an algorithm for tableau construction for BDD-based representation. (The description in Section 3.4.2 follows this approach.) Their complexity results are comparable to those of symbolic CTL model checking (see the comparative case study in [CGH97]).

Secondly, the length of an LTL formula is much shorter than the comparable CTL formula in some cases. It can be shown that for some LTL formulas the corresponding CTL formula has exponential size. Moreover, if it is possible to generate an LTL formula from a CTL formula it will be in most case shorter, but never longer then the branching-time version. Thus, the polynomial complexity of CTL is a relative advantage only.

3.5.2   Expressiveness of Temporal Logics

In [Lam80], Lamport compared the expressiveness of branching-time versus linear-time logics for the first time. He showed that each logic could express certain properties that could not be expressed in the other. However, his resulting notion of *expressive equivalence* failed because it classified some satisfiable formulas as equivalent to false. The comparison lacked a suitable framework since formulas of linear time are related to individual paths and formulas of branching-time are related to states.

Emerson and Halpern [EH86] introduced a uniform logical framework for comparing the expressive power of both logics called CTL*. Within CTL* linear-time as well as branching-time properties can be expressed and the formulas of both logics are related to states.

The proofs of the inexpressibility results in [EH86] are very long and complicated. [CD88] introduced instead a simple characterisation of those CTL* formulas that can be expressed in LTL. Also they gave a necessary condition that a CTL* formula must satisfy in order to be expressible in CTL.

In the following subsection, we will only sketch the relation between the logics and will show some differences for the expressiveness by means of examples.

The Extension CTL*

Besides the usual first-order operations, CTL* provides two different kinds of *temporal operators*. Linear temporal operators stating propositions with respect to a path. Basically, we may use the next operator **X** and the until operator **U**. Temporal operators of branching-time referring to several branching paths that start in the current state. A proposition must hold for some paths **E**, or for all paths **A** in the computation tree. These operators can be found in the sub-logics as well (compare Section 3.3.1 and Section 3.4.1). Again we distinguish state and path formulas.

Let $\mathcal{V}$ be the set of atomic propositions, then the syntax of CTL* is defined in the following way:

Definition 3.21:   Syntax of CTL*

    state formulas:

        i.) If $\varphi \in \mathcal{V}$, then φ is a state formula.

        ii.) If φ and ψ are state formulas, then $\neg\varphi$ and $\varphi\vee\psi$ are state formulas.

        iii.) If φ is a path formula, then $\mathbf{E}\,\varphi$ is a state formula (the same holds for $\mathbf{A}\,\varphi$).

    path formulas:

        i.) If φ is a state formula, then φ is also a path formula.

        ii.) If φ and ψ are path formulas,
        then $\neg\varphi$, $\varphi \vee \psi$, $\mathbf{X}\,\varphi$ and $\varphi \,\mathbf{U}\,\psi$ are path formulas.

The relation between state and path formula in CTL* is depicted in Figure 3.5.

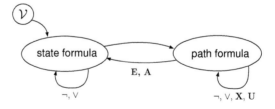

Figure 3.5: State formulas and path formulas in CTL*

CTL versus CTL*

The major difference to CTL is that in CTL* every state formula is a path formula as well. Thus, in CTL* *every* formula can be combined with the quantifying operators over paths ($\mathbf{E}$ and $\mathbf{A}$) which does not hold for CTL. (Recall from definition 3.15 that only if φ is a path formula then $\mathbf{E}\,\varphi$ is syntactically correct.) For example, the formula $\mathbf{E}\,(\varphi \wedge \mathbf{X}\,\psi)$ is syntactically correct in CTL* but not in CTL. In CTL every state formula has to be preceded by a path quantifier.

Compared to CTL* (see Figure 3.5) we get the relation between state and path formulas within the CTL framework as sketched in Figure 3.6.

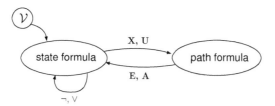

Figure 3.6: State formulas and path formulas in CTL

Fairness.   Accordingly, in CTL the sub-formula $\mathbf{G}\,\varphi$ needs to be quantified over paths. We cannot express $\mathbf{F}\,\mathbf{G}\,\varphi$, for instance. The CTL formula $\psi = \mathbf{AF}\,(\mathbf{AG}\,\varphi)$ has a different semantics. The structure M, depicted in Figure 3.7, gives an example: Although the LTL formula $\psi' = \mathbf{F}\,\mathbf{G}\,\varphi$ holds in M, i.e., $M, s_0 \models \mathbf{A}\,(\mathbf{F}\,\mathbf{G}\,\varphi)$, this is not the case for the CTL formula $\psi = \mathbf{AF}\,(\mathbf{AG}\,\varphi)$, i.e., $M, s_0 \not\models \mathbf{AF}\,(\mathbf{AG}\,\varphi)$. As an explanation, one might consider that in the computation tree of M the path for the computation that stays in state s_0 has in every node a successor that does not satisfy φ. At every state it is possible to transit to state s_1. That is, there is no point in the future at which φ is always true in every possible future.

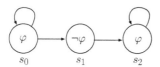

Figure 3.7: Branching-time structure M

The same argument holds for conditional fairness which is useful in many applications. We distinguish *weak fairness* and *strong fairness*.

Weak fairness (also known as *justice*) (which is the most common notion of fairness) is defined as an LTL formula which has no counterpart in CTL. An infinite path is *weak fair* with respect to a set of fairness conditions $\Xi = \{\xi_1, \xi_2, \ldots\}$ and property φ if it satisfies the following: if each $\xi_i \in \Xi$ eventually holds forever then it follows that φ holds infinitely often during the path. This can be formalised as follows:

$$(\textstyle\bigwedge_{i=1}^{n} \mathbf{F}\,\mathbf{G}\,\xi_i) \implies \mathbf{G}\,\mathbf{F}\,\varphi \tag{3.23}$$

Also strong fairness cannot be expressed in CTL. An infinite path is *strongly fair* with respect to the fairness conditions in Ξ and property φ in the following case: if each $\xi_i \in \Xi$ holds infinitely often, then it follows that φ holds infinitely often along the path.

$$(\bigwedge_{i=1}^{n} \mathbf{G} \, \mathbf{F} \, \xi_i) \implies \mathbf{G} \, \mathbf{F} \, \varphi \tag{3.24}$$

For a proof of these results see [EH86].

LTL versus CTL*

LTL, on the other hand, lacks the path quantifiers $\mathbf{E}$ and $\mathbf{A}$. The semantics of LTL formulas is given with respect to paths, not to states. State formulas are not expressible. This is shown in Figure 3.8. In the framework of CTL*, however, every LTL formula φ is regarded as implicitly quantified over all paths. We consider only formulas of the form $\mathbf{A} \, \varphi$. But an LTL formula cannot express that something is satisfied along *some* computation path, which is expressible in branching-time logic.

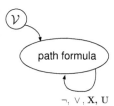

Figure 3.8: Formulas in LTL

Obviously, fairness is expressible in LTL since there is no restriction for combining the temporal operators (in fact, the formulas given in 3.23 and 3.24 are LTL formulas).

Besides fairness, we are also interested in using LTL for specifying environment behaviour. As shown in Section 5.1, specifying assumptions on environment behaviour in a constructive way by means of ASM transition rules is quite difficult. The use of a logical framework for this concern would be helpful. LTL formulas model a behaviour of a component in an intuitive way by means of particular executions, i.e., state sequences or paths. For instance, *"if a request arrives then eventually it will be acknowledged"* can be modelled by $(\neg req \wedge \mathbf{X} \, req \rightarrow \mathbf{F} \, ack)$. Such a sequence of changes (e.g., the *arrival* of a request) and following executions cannot easily be modelled with CTL.

Intersections of the logics

Comparing the expressiveness of the three different temporal logics CTL*, CTL and LTL, it can be shown that CTL* comprises both other logics. The intersection of LTL and CTL is a non empty set of formulas. However, not every LTL formula is expressible in CTL and vice versa. The overall framework is depicted in Figure 3.9. Some examples are given for the corresponding sub-sets.

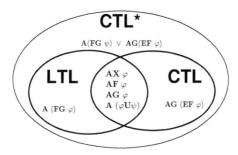

Figure 3.9: Expressiveness of the three temporal logics

This figure concludes the comparison. It should be clear that both logics, for branching time and linear time, have their own advantages and disadvantages. In fact, a combination of both logics appears promising. Model checking for CTL* that comprises both logics is a very complex task and therefore in many cases not feasible. It is recommended to restrict the treated logic to the actual needs. One approach is described in Section 3.4. It combines model checking for CTL with a treatment for LTL.

In the next chapter, we describe the data structure used for symbolic model checking: *binary decision diagrams* and its extension *multiway decision graphs*. The latter one provides the target language for our interface from ASM to the MDG-Package. A basic understanding of this data structure should help to follow the description of the interface given in Chapter 8.

Chapter 4

Decision Graphs

The complexity of a model checking algorithm can be reduced by using a symbolic representation of states and state transitions. This approach is called *symbolic model checking* (c.f., section 3.3.3 on page 34). The representation is based on the idea of describing a system by means of boolean functions. Boolean logic is appropriate for specifying hardware circuits. Model checking was introduced as an automatic approach of hardware verification.

Boolean functions can be canonically represented by *binary decision diagrams* (BDDs). For this graph structure, very efficient algorithms are available which provide the basis for model checking (e.g., [Bry86] or [DB98a]). For more complex systems, involving higher-level data structures (not only boolean values), a coding is necessary that maps these data onto boolean values. To avoid this coding, alot of research is being done in order to lift the boolean graphs to more complex data domains depending on the application domain that is considered; for instance "multi-values" for modelling arbitrary sets, real numbers for coding time, intervals, etc. For specifying hardware systems, the notion of "word-level" is used to name a higher level of abstraction for the system specification (e.g., multiplication modules are easier to describe on the word-level rather than the boolean level). [DB98b] gives an overview of various types of decision diagrams and develops their common theory.

In this Chapter, we give a short overview of the notion of BDDs (Section 4.1), which are the most common symbolic representation used for model checking. In Section 4.2 we introduce a special sort of lifting of this simple structure, called *multiway decision graphs* (MDGs). MDGs are a kind of superset of BDDs which provide a higher level for the representation. For our work, both data structures are relevant since they are internally used in the model checking tools we want to apply to ASM.

4.1 Binary Decision Diagrams

Syntactically, a decision diagram (DD) is defined as a finite, connected, directed acyclic graph $G = (V, E)$ over a set of Boolean variables X_n and a non empty terminal set T; V denotes the set of vertices (nodes) and E the set of edges. The non-terminal nodes in V are labelled with variables from X_n. They have two successor nodes and the two edges leading to them are labelled with 0 (*low*) and 1 (*high*), respectively. The successor node of v that is connected via the "low" edge is denoted as $low(v)$, and the successor node connected via the "high" edge is denoted as $high(v)$. The terminal nodes are labelled with values from terminal set T. The topmost node is called the *root* of the graph. Such a DD G may have the following properties:

- G is *complete*, if each variable in X_n is encountered exactly once on each path in G from the root to the terminal node.

- G is *free*, if each variable in X_n is encountered at most once in each path.

- G is *ordered*, if it is free and the variables on each path are encountered in the same order.

A *binary decision diagram* (BDD) is a DD with the terminal set $T = \{0,1\}$ which represents the truth values of boolean logic. Accordingly, a BDD represents a boolean function. We distinguish two cases for a diagram B:

- If B consists of a single terminal node, then it represents the truth value the node is labelled with, i.e., 0 or 1.

- If B has a root v labelled with a variable $x_i \in X_n$, the low edge leads to a subgraph that represents the formula $f_{low(v)}$ and the high edge leads to a subgraph that represents $f_{high(v)}$ then B represents the formula f_v over variables in X_n

$$f_v(x_1, \ldots, x_n) = (\bar{x}_i \wedge f_{low(v)}(x_1, \ldots, x_n)) \vee (x_i \wedge f_{high(v)}(x_1, \ldots, x_n)) \quad (4.1)$$

This definition of the semantics of a BDD is also known as the *Shannon decomposition* (c.f., [DB98a]).

Figure 4.1 gives an example. The BDD represents the function $f_v(x_1, y_1, x_2, y_2) = (x_1 \Leftrightarrow y_1) \wedge (x_2 \Leftrightarrow y_2)$. The dashed lines denote the low edges and the solid lines denote the high edges. The label of each terminal nodes represents the evaluation of the path that leads to it, i.e., either 0 or 1. This graph is complete and ordered.

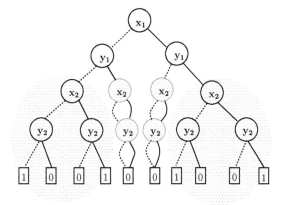

Figure 4.1: Complete OBDD that represents the function $(x_1 \Leftrightarrow y_1) \wedge (x_2 \Leftrightarrow y_2)$

Ordered BDDs (OBDDs) can be reduced if they contain (a) redundant nodes whose edges lead to the same
successor node or (b) isomorphic subgraphs. The OBDD
in Figure 4.1, for instance, contains redundant nodes labelled with x_2 and y_2 (these nodes are shaded in the
figure). Their evaluation does not change the value of
the function since both edges lead to the same successor.
Furthermore, the two subgraphs with root x_2 (marked by
a patterned background) are isomorphic. We can reduce
a complete OBDD by erasing the redundant nodes and
merging the isomorphic subgraphs. Figure 4.2 shows the
reduced OBDD (ROBDD) after applying the reduction
algorithm. Note that the order of the variables, here
$x_1 < y_1 < x_2 < y_2$, essentially influences the size of the
ROBDD. A changed order of $x_1 < x_2 < y_1 < y_2$ would
lead to an ROBDD with nine nodes. The problem of
finding the optimal order for an arbitrary graph is NP

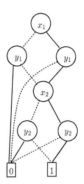

Figure 4.2: Reduced OBDD

hard ([Bry86]). However, heuristics are available that provide in most cases an appropriate solution. Only in some cases, the ROBDD representation of a function
grows larger than other representations of the same function. The most famous example is the function of an integer multiplier (see [Bry86]) which grows exponentially
for any variable order. Since an ROBDD is a canonical representation, satisfiability
and equivalence are easy to check: A function is satisfiable if the corresponding
graph is not the terminal node labelled with 0. Two functions are equivalent if they
are represented by the same ROBDD.

The importance of ROBDD for model checking is given through the fact that
sets and relations can be modelled by boolean characteristic functions. That is,
transition systems, comprising a set of initial states and a state transition relation,
can be represented by ROBDDs. This *symbolic representation* is in most cases
much smaller than other state representations. The complexity of symbolic model
checking algorithm depend on the size of the ROBDD rather than the actual state
space, and is therefore more efficient in cases where the symbolic representation by
ROBDDs is smaller than the explicit representation.

4.1.1 Algorithms for ROBDDs

Very efficient algorithms were developed that realise the boolean operation for ROB-
DDs (see [Bry86] or [DB98a]). For symbolic model checking of CTL formulas (see
Section 3.3 on page 28) another algorithm for ROBDDs is central, called the *relational product* (introduced in [BCL+94]). In the following we describe this algorithm
in more detail to aid understanding of the further development based on MDGs.

The algorithm for relational product implements the computation of `CheckEX`
(equation 3.10 in Section 3.3.3 on page 34):

$$\text{CheckEX}(\Phi(\vec{s})) := (\exists \vec{s}'.(\Phi(\vec{s}') \land R(\vec{s}, \vec{s}')))$$

Accordingly, the algorithm realises the existential quantification over primed state
variables $\vec{s}'$ and the conjunction of two ROBDDs that represent the boolean formulas $\Phi(\vec{s})$ and $R(\vec{s}, \vec{s}')$.

We illustrate the algorithm in Figure 4.3. The input parameters B and G denote
the BDDs that represents $\Phi(\vec{s}')$ and $R(\vec{s}, \vec{s}')$. The set of variables *Var* corresponds
to the primed variables in $\vec{s}'$ as given above in the formula for `CheckEX`. They are
existentially quantified.

The algorithm is defined recursively on the structure of the BDDs. That is, we
choose among the root node labels of both graphs the one that comes first in the

---

Bin_RelP $(B, G : \text{ROBDD}, \text{Var} : \text{set of variables}) =$

  if $B = \text{false}$ or $G = \text{false}$
    then return *false*
  else if $B = \text{true}$ and $G = \text{true}$
    then return *true*
  else if Bin_RelP$(B, G, \text{Var}) = R$ is already computed
    then return R

  else
    let x be the root label of B
    let y be the root label of G
    let z be the topmost of x and y

    $R_0 \;:=\; \text{Bin\_RelP}(B \mid_{z=0}, G \mid_{z=0}, \text{Var})$
    $R_1 \;:=\; \text{Bin\_RelP}(B \mid_{z=1}, G \mid_{z=1}, \text{Var})$

    if $z \in \text{Var}$
      $R \;:=\; \mathbf{Or}(R_0, R_1)$
    else
      $R \;:=\; (z = 1 \;\wedge\; R_1) \vee (z = 0 \;\wedge\; R_0)$
    endif

    insert (B, G, Var, R) in the result cache

    return R
  endif

end Bin_RelP

Figure 4.3: The algorithm for relational product on ROBDDs

order of variables, denoted by z in the figure. Two cases may occur: (a) z appears in both graphs as the root node label; (b) only one graph contains z and the other graph does not. The subgraphs of a BDD are computed by substituting z with the values 0 and 1; we get $F \mid_{z=0}$ and $F \mid_{z=1}$, where $F \in \{B, G\}$. If F does not contain the root label z then the subgraphs are equal to F, i.e., $F \mid_{z=0} = F \mid_{z=1} = F$.

We apply the relational product algorithm, Bin_RelP, to these subgraphs. The resulting graphs, called R_0 and R_1 in Figure 4.3, are used as parameters when computing the relational product at the level of root node z:

- If z is an element of the quantified set of variables, i.e., $z \in \text{Var}$, then we disjoin both subgraphs R_0 and R_1 by applying the **Or** algorithm on BDDs where disjunction realises the existential quantification (c.f., the definition of CheckEX above). Due to the quantification, z does not occur in the resulting graph R any more.

- If $z \notin \text{Var}$ we build a BDD R that represents the formula $(z = 1 \wedge R_1) \vee (z = 0 \wedge R_0)$ by keeping z as the root node label and using R_1 and R_0 as high and low subgraphs.

Besides the relational product, we use algorithms for conjunction and disjunction of ROBDDs. A more general algorithm Apply, which takes the boolean operation to be applied as a parameter, is described in detail in the literature (see [DB98a]). These simple algorithms together with fixpoint algorithms are sufficient for the symbolic model checking approach (c.f., Section 3.3).

4.2   Multiway Decision Graphs

Multiway decision graphs (MDGs) are a generalisation of ROBDDs. They allow symbolic representation of sets of states and transition relations in the same way as the binary graphs. Additionally, they incorporate variables of any enumerated sort and even abstract sorts. This essentially lifts the level of abstraction of the model coding. Firstly, the binary coding of (finite) data types, which yields more complex representations, can be avoided. Secondly, even infinite data types can be represented and by means of *abstract sorts* and *uninterpreted functions* incorporated into the model checking process.

Lifting the low-level coding of BDDs is also used elsewhere. For example, *multiway decision diagrams* (MDDs), used as an internal data structure in the model checker VIS ([Gro96]), allow for variables of enumerated sort as well. However, they do not include abstract data types.

MDGs are introduced in the literature in [CZS+97] and [CCL+97]. Several examples show that model checking based on MDGs is successfully used for verifying hardware circuits (see for example [ZST+96], [BT98], or [LTZ+96]). We want to use this approach for model checking ASM (see Chapter 8) and therefore recall here the basic notion of this graph structure and the logic it is based on.

4.2.1   The Logic

In the same way that BDDs represent formulas of boolean logic, MDGs are a canonic representation of formulas of a special many-sorted first-order logic. The logic comprises various sorts, constants, variables, and function symbols. Constants and variables have a sort, and a function symbol of arity n has a type $\alpha_1 \times \ldots \times \alpha_n \to \alpha_{n+1}$ where the α_i are sorts.

In contrast to ordinary many-sorted first-order logic, this logic distinguishes between *concrete* sorts and *abstract sorts*. Concrete sorts have enumerations, abstract sorts do not. This leads to the distinction between three kinds of functions: Assume that f is of type $\alpha_1 \times \ldots \times \alpha_n \to \alpha_{n+1}$, then

- f is an *abstract function symbol*, if α_{n+1} is an abstract sort;

- f is a *concrete function symbol*, if all α_i are concrete sorts;

- f is a *cross operator*, if α_{n+1} is a concrete sort and at least one of $\alpha_1, \ldots, \alpha_n$ is an abstract sort.

The elements of an enumeration are called *individual constants*. Constants of abstract sort are denoted as *generic constants*.

Semantics

An *interpretation* is a mapping ψ that assigns a denotation to each sort, constant and function symbol. It satisfies the following conditions:

1. The denotation of an abstract sort α is a non-empty set $\psi(\alpha)$.

2. For each concrete sort $\alpha = \{a_1, \ldots, a_n\}$ the denotation is given as $\psi(\alpha) = \{\psi(a_1), \ldots, \psi(a_n)\}$.

3. If f is a function symbol of type $\alpha_1 \times \ldots \times \alpha_n \to \alpha_{n+1}$ then $\psi(f)$ is a function of type $\psi(\alpha_1) \times \ldots \times \psi(\alpha_n) \to \psi(\alpha_{n+1})$.

A *variable assignment* is a mapping that assigns a constant value to each variable in X, the set of variables. A variable assignment ϕ is called ψ-*compatible* if it maps each variable $x \in X$ of sort α to value $\phi(x) \in \psi(\alpha)$. Φ_X^ψ denotes the set of ψ-compatible variable assignments for the variable set X. A formula P is true (or false) with respect to the interpretation and the variable assignment. We write $\psi, \phi \models P$ if P denotes truth under interpretation ψ and ψ-compatible variable assignment ϕ. Abstract function symbols and cross-operators are *uninterpreted*.

Atomic formulas are the truth values, **true** and **false**, and equations. A well typed *equation* is an expression $A_1 = A_2$ where the left hand side (LHS) and the right hand side (RHS) are of the same sort. *Formulas* are built from atomic formulas using logical connectives ($\vee, \wedge, \neg, \ldots$) and quantifiers. This logic is similar to the logic that is used to describe the states of an ASM (c.f., Section 2.2), except for abstract sorts and uninterpreted functions.

MDGs represent a subset of all formulas which is sufficient for encoding transition systems. [CCL+97] introduces the notion of *directed formulas* to capture this subset of formulas.

Directed Formulas

A *directed formula* (DF) is given in terms of two disjoint sets of variables U and V. Set V can be seen as the set of dependent variables and U as the set of independent variables. Both sets comprise variables of concrete and abstract sort, i.e., $U = U_{conc} \cup U_{abs}$ and $V = V_{conc} \cup V_{abs}$ where the set index indicates the kind of sorts of its elements. In a DF, the variables in V_{abs} may only appear as a left hand side (LHS) of an equation. Every abstract variable occurring on the right hand side (RHS) of an equation or as a parameter of a cross-term is an element of U_{abs}. The LHS of an equation of concrete sort (if not a cross-term) may be variables of either sets U or V. U and V determine the *type* of a DF, which is denoted as $U \rightarrow V$.

A DF of type $U \rightarrow V$ is a formula in disjunctive normal form (DNF) that satisfies the following three conditions:

(i.) Each disjunct is a conjunct of equations of one of the following forms

   $f(B_1, \ldots, B_n) = a$
   where f is of concrete sort α and at least one of the terms B_i is of abstract sort (i.e., f is a cross-operator), and all variables in any term B_i are elements of U; a is an individual constant in the enumeration of sort α;

   $w = a$
   with $w \in (U \cup V)$ is a variable of concrete sort α and a is an individual constant of the same sort;

   $v = A$
   with $v \in V$ is a variable of abstract sort α and A is a term of sort α containing only variables that are elements of U.

(ii.) In each disjunct, the LHSs of the equations are pairwise distinct.

(iii.) Every abstract variable $v \in V$ appears as the LHS of an equation $v = A$ in each of the disjuncts.

A DF is said to be *concretely reduced* if and only if every term A in an equation $v = A$ is a concretely reduced term. A term is called concretely reduced if it contains no concrete terms other than individual constants. That is, in a concretely reduced term every concrete variable and every function of concrete sort (this includes cross-terms) has to be substituted by its value.

A variable has *primary occurrence* in a DF if it occurs as the LHS of an equation in the DF. A variable has *secondary occurrence* in a DF if it occurs either in a cross-term on the LHS of an equation, or it occurs in a concretely reduced term on the RHS of an equation.

4.2.2 Multiway Decision Graphs

An MDG is a finite graph $G = (V, E)$, whose non-terminal nodes are labelled by terms and whose edges, starting from a non-terminal node, are labelled by terms of the same sort as the node. Terminal nodes are labelled by formulas. Generally, a graph G represents a formula in the following way:

- If G consists of a single terminal node, then it represents the formula the node is labelled with.

- If G has a root node labelled with term A and edges labelled with terms $B_1, \ldots, B_n$ leading to subgraphs $G_1, \ldots, G_n$ that represent formulas $P_1, \ldots, P_n$, then G represents the formula

$$(A = B_1 \wedge P_1) \vee (A = B_2 \wedge P_2) \vee \ldots \vee (A = B_n \wedge P_n). \tag{4.2}$$

This definition is similar to the definition of BDD representation given by Equation (4.1). In contrast, however, the number of edges of an MDG is not restricted to two and the labels of nodes and edges do not have to be boolean variables or values. In Figure 4.4 we depict Formula (4.2) as a graph G.

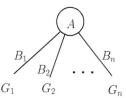

Figure 4.4: The MDG G

To be a canonical representation, an MDG has to satisfy certain *well-formedness conditions*. These conditions involve two kinds of orders: (a) The *standard term order* is a total order on the set of all terms of the logic. (b) The *custom symbol order* is a total order on the set of symbols C which includes all cross-operators, all concrete variables, and some of the abstract variables. It is given by the user and need not be compatible with the standard term order. The custom symbol order is a generalisation of the variable order for ROBDDs.

MDG Well-formedness Conditions

Assume U and V are disjoint sets of variables such that all abstract variables in V participate in the custom symbol order. An MDG of type $U \to V$ is a directed acyclic graph with one root and ordered edges, that represents a *concretely reduced* DF. The following *well-formedness conditions* have to be satisfied:

1. Every leaf node is labelled with the formula **true**, except when G consists of only one terminal node which may be labelled with **true** or **false**.

2. Every node-edge pair (N, E), where N is an internal node, represents an equation of the DF, such that N is labelled by its LHS and E by the RHS. The LHS and RHS follow the definition of DFs formulated above. Each path in the graph describes one of the conjuncts of equations of the represented DF. The whole graph represents the disjunction of all conjuncts of the DF.

3. The graph is ordered. All edge labels of one node are ordered according to the standard term order[1]. Along each path, all variables and cross-operators of cross-terms that label the nodes, are ordered by the custom symbol order. If cross-operators appear more than once among the labels of one path, the order of the corresponding cross-terms is given through the standard term order. The order of nodes and edges must be total; repeated labels are not allowed.

4. The MDG must be minimal. Analogously to the reduction for OBDDs, MDGs can be reduced by merging isomorphic subgraphs, and deleting redundant nodes. In the case of MDGs, a node is redundant if it is labelled by a concrete variable or cross-term, the labels of the outgoing edges comprise the enumeration of its sort entirely, and all edges lead to the same subgraph.

5. No variable should have both, primary and secondary occurrences in the same graph. This is guaranteed if U and V are disjoint sets and the MDG is of type $U \to V$ and satisfies condition 2.

6. The set of abstract variables having primary occurrences along a path is the same for all paths in a given graph. This is satisfied if the graph represents a DF according to condition 2.

7. The order of abstract variables must consider the dependencies. Dependent abstract variables must have a lower order than those abstract variables they depend on. Two cases can occur: (a) if the node-edge pair represents the equation $w = A(v)$ where w is an abstract variable and A a term that contains the abstract variable $v \in V$, then the custom symbol order must satisfy $v < w$; (b) if a node is labelled by the cross-term $f(B_1, \ldots, B_n)$ and the abstract variable $v \in V$ occurs in one of the parameter terms B_i and participates in the custom symbol order, then the custom symbol order has to satisfy $v < f$.

   This condition is due to the fact that in both cases the abstract variable $v \in V$ must be substituted during the computation on the graph.

Given a custom symbol order and a standard term order, it can be proven for two well-formed MDGs G and G' representing formulas P and P', respectively, that if $\models P \Leftrightarrow P'$, then G and G' are isomorphic (for a proof see [CZS+97]). That is, MDGs are a canonical representation for DFs. Throughout this work we use the notion MDG for denoting a well-formed MDG.

4.2.3 Operations on MDGs

Similar to ROBDDs, operations on MDGS can be implemented in a very efficient way. The MDG-package yields a library of the basic functions implemented prototypically in Prolog. This implementation lacks the efficiency of C/C++ code (due to the extensive use of memory space in Prolog which influences the runtime). However, the algorithms are in principle as efficient as the corresponding algorithms on ROBDDs (see [ZST+96] for a comparison of different tools). [Zho96] provides a very helpful documentation of this package, which allows to develop more complex algorithms based on MDGs, as for instance algorithms for model checking temporal logic formulas (see [XCS+98, Xu99]).

In the following subsection, we explain the basic functions on MDGs that are useful for model checking. All functions process lists of MDGs. For the application, we have to follow particular restrictions in order to preserve the well-formedness conditions for the resulting graph.

---

[1] In fact, edge labels and corresponding subgraphs are encoded as lists. These lists are ordered.

Disjunction

The disjunction of two graphs can be easily computed in terms of the disjunction of the subgraphs if both graphs have the same root label. Figure 4.5 sketches the situation for a root label A (of abstract or concrete sort). In the result, the subgraph G' represents the disjunction $(G_2 \vee G_4)$ and G'' stands for $(G_3 \vee G_5)$.

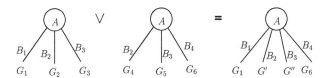

Figure 4.5: Disjunction of two MDGs with the same root label

If the graphs have different root labels, however, it is not always possible to compute the disjunction. For instance, if the root labels are abstract variables v and u, and the graphs represent the logical formulas $(v = a)$ and $(u = b)$, then there is no well-formed MDG that represents their disjunction. The formula $(v = a) \vee (u = b)$ does not satisfy the condition (iii.) on DFs since not every abstract variable appears as an LHS of an equation in every path.

If node label v is of concrete sort and it comes before u in the order of nodes, then we may add a redundant node on top of the second MDG and thus get the situation for identical root labels. This redundant node must comprise all values of its enumeration as labels of outgoing edges, and these edges must lead to the same subgraph.

In order to apply disjunction, we must assure that both MDGs must have the same set of primary abstract variables, and, therefore, if the root labels are abstract they are the same variables (c.f., condition (iii.) on DFs). This restriction also motivates the well-formedness condition 6. on MDGs given in the list above.

The disjunction algorithm of the MDG-package is n-ary. It takes a list of MDGs P_i of type $U_i \rightarrow V$ as input and produces the disjunct $(\bigvee_{1 \leq i \leq n} P_i)$ of type $(\bigcup_{1 \leq i \leq n} U_i) \rightarrow V$.

Relational Product

The relational product computes the conjunct of MDGs and existentially quantifies over a given set of variables, similar to the relational product on ROBDDs (see function **Bin_relP** in Figure 4.3). As an extension the algorithm on MDGs processes lists of MDGs rather than two graphs only. This allows *partitioning of the transition relation* (i.e., the relation is represented by a set of smaller MDGs rather than by a single big graph) and therefore the algorithm benefits from *early quantification of variables* (for more details see [BCL+94]).

However, the application on MDGs is restricted in the following sense: Generally, two MDGs with roots that are labelled with the same abstract variable can not be conjoined. For instance, let u be an abstract variables, and a and b be two distinct abstract generic constants, then we cannot compute the conjunction $(u = a) \wedge (u = b)$. It is undecidable if this formula is true or false, since the abstract generic constants are uninterpreted; they can be equal or not. Thus, the relational product can be computed only if the sets of primary abstract variables of the MDGs to be conjoined are disjoint.

The algorithm for computing the relational product on MDGs combines conjunction, existential quantification, and renaming. It takes as input n graphs P_i

of type $U_i \to V_i$, the set of variables E to be quantified, and a substitution function η for renaming and yields as the result the graph that represents the formula $\left((\exists v \in E)(\bigwedge_{1 \leq i \leq n} P_i) \cdot \eta\right)$. For $1 \leq i < j \leq n$, the variable sets V_i and V_j must not have any abstract variable in common. The type of the resulting subgraph is $\left((\bigcup_{1 \leq i \leq n} U_i) \backslash (\bigcup_{1 \leq i \leq n} V_i)\right) \to \left(((\bigcup_{1 \leq i \leq n} V_i) \backslash E) \cdot \eta\right)$. The algorithm is documented in detail in the appendix of [CZS$^+$97].

Pruning by Subsumption

The algorithm for pruning by subsumption is useful for approximating the difference of two graphs. The input is two graphs P and Q of types $U \to V_1$ and $U \to V_2$, where U contains only abstract variables that do not participate in the custom symbol order. The result graph is of type $U \to V_1$ and is derivable from P by *pruning*.

Pruning means removing those paths from P that are subsumed by graph Q. A path in P is subsumed if there is a substitution on variables in U such that applied to a path in Q this new path contains all node-edge pairs of the subsumed path in P. The resulting graph R satisfies the following condition

$$P \wedge \neg (\exists u \in U)Q \Rightarrow R \qquad (4.3)$$

Since R is computed by pruning of P the paths of R represent a subset of disjuncts represented by the paths in P, i.e., $R \Rightarrow P$. Together with the logical implication in (4.3) we can view R as approximating the logical difference between P and $(\exists u \in U)Q$. An equivalent form of this formula cannot be found in general. If R is **false**, it follows that $P \Rightarrow (\exists u \in U)Q$.

Negation

The negation of an MDG can be computed only if it contains no primary abstract variables. If u is an abstract variable and a is a generic abstract constant, then we cannot evaluate the formula $\neg(u = a)$, since we have no interpretation for u and a. In contrast to the negation, the equation $(u = a)$ implicitly represents the existential quantification of a, i.e., there exists a value a such that u equals a. However, this does not include the knowledge of a witness that satisfies this equation. Therefore, we cannot express that u is *not* equal the generic abstract constant a.

For MDGs with all primary variables of concrete sort, we can compute the negation.

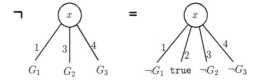

Figure 4.6: Negation of MDG with node label of concrete sort

See for an example Figure 4.6. We assume $x \in X$ and $X = \{1, 2, 3, 4\}$. The left MDG in the figure represents the formula

$$\neg\big((x = 1 \wedge G_1) \vee (x = 3 \wedge G_2) \vee (x = 4 \wedge G_3)\big)$$

$$
\begin{aligned}
= \ & \neg(x = 1 \wedge G_1) \wedge \neg(x = 3 \wedge G_2) \wedge \neg(x = 4 \wedge G_3) \\
= \ & \big((x \neq 1) \vee (x = 1 \wedge \neg G_1)\big) \\
& \wedge \big((x \neq 3) \vee (x = 3 \wedge \neg G_2)\big) \wedge \big((x \neq 4) \vee (x = 4 \wedge \neg G_3)\big) \\
= \ & (x \neq 1 \wedge x \neq 3 \wedge x \neq 4) \vee (x = 1 \wedge \neg G_1) \vee (x = 3 \wedge \neg G_2) \vee (x = 4 \wedge \neg G_3) \\
= \ & (x = 2) \vee (x = 1 \wedge \neg G_1) \vee (x = 3 \wedge \neg G_2) \vee (x = 4 \wedge \neg G_3)
\end{aligned}
$$

The result formula is represented by the right MDG in Figure 4.6. This example provides a general approach to computing the negation of any MDG that contains only primary variables of a concrete sort which can be enumerated.

In this thesis, we implement an algorithm for negation, which is not given in the MDG-package.

With this chapter, we finish the second part of the thesis which presents the theory of model checking and the underlying data structure that is exploited for symbolic model checking. In the next chapter, we introduce our approach for model checking ASM. Technically, the central task of this approach is a transformation between languages. Apart from this transformation, the application of model checking needs some insights into treating the models. The next chapter discusses the intricacies of applying model checking to ASM and introduces the core transformation algorithm. The correctness proof of the transformation algorithm completes the chapter.

Chapter 5

Applying Model Checking to Abstract State Machines

An ASM model of a system comprises the state space, defined by means of universes and functions, and the state transition behaviour, defined by means of transition rules. Thus, an ASM model is basically a transition system as used within the model checking approach (c.f., Section 3.2.1 on page 24). In principle, model checking can be applied.

We understand model checking as tool support for *debugging* an ASM model. Debugging is a cyclic process of repeatedly reworking the model according to errors that might be found. A debugging cycle for ASM models that involves model checking is depicted in Figure 5.1: For applying model checking, the ASM model has to be transformed into the input language of the chosen model checker. In Figure 5.1, the model in input language of the model checker is denoted as MC model. Once a model is transformed, we benefit from fully automatically checking safety and liveness properties which we call the *requirements*. If the model checker outputs a *counterexample* the user has to inspect the erroneous behaviour and correct the ASM model correspondingly. For complex systems, a counterexample yields a good insight into the behaviour of the model. It is given as a sequence of states that lead to a state in which the given requirement is violated. This sequence of states can be fed into the simulation tool for ASM and thus triggers a simulation run that shows the erroneous behaviour of the model (see discussion in Section 6). The simulation of the counterexample is depicted by dotted lines reflecting that this interface is not yet implemented.

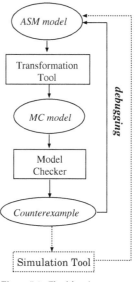

Figure 5.1: The debugging process

The transformation and checking procedure is implemented and runs fully automatically. When the ASM model is changed in order to correct an error, the process can be repeated. This repetitive *debugging process* works nicely if the transformation is fast, and the checking procedure provides results in a reasonable span of time, where the term "reasonable" depends on the personal taste of the user.

Making use of this approach, practical experience points out the limitations when using particular checking tools. We also gain further insight of what has to be considered when applying model checking to ASM models.

Limitations in the size of the state space. Firstly, the state space of the model has to be finite and small enough. Although it might be easy to transform ASM models with a huge state space, the resulting model might be too big for the model checker. If the user has to wait for a response of the checker over several days, the debugging process can hardly be followed. (Sometimes, even waiting for some hours can be too long.) Thus, it is prudent to keep the model small enough. A means of keeping a model small is *abstraction*.

In our first approach (see Chapter 6 on page 83), we use the SMV model checker ([McM93]), which gives no support for abstraction. We simply rely on finding a suitable ASM model. In Section 2.1.2 on page 10, we introduced the practice of modelling ASM on different levels of abstraction. Among those different layers we might find a model that is small enough for being model checked – but on the other hand big enough (in terms of being concrete) for expressing the properties we want to check. However, this solution is not convincing. The process of modelling should not be influenced by the later process of model checking. A level of abstraction should not be chosen with respect to the resulting state space. Instead, modelling should focus on a clear and succinct specification of the problem at hand. Models that are succinct in terms of their textual representation and easy to understand do not necessarily have a small state space; they may even have infinite state spaces.

In our second approach (see Chapter 4.2), we interface the ASM Workbench with the MDG package ([CZS+97], [CCL+97]). This tool set supports abstraction by means of its internal data structure that comprises abstract data types. We show how to treat infinite models by introducing abstract types and functions into the ASM specification language. The resulting abstract models can be seen as infinite models. They can be checked within the MDG framework.

The state space has to be fixed. We call the state space of the model *fixed* if the universes are not extended during a run. For model checkers, we have to declare the domains of the state space before running the checking procedure. If we would consider domains to be extendible they would be regarded as possibly infinite. Thus, the state space can not be fully explored. It is generally impossible to treat "open" domains that can be extended. In ASM, dynamic extension of universes (or domains) can be modelled using the **import** rule (c.f., rule 2.7 on page 18 and 2.13). Therefore, the **import** rule is not treated by our transformation.

A different situation is given by introducing the notion of abstract types into the ASM language when using the MDG approach. Domains of abstract type do not require an interpretation, and thus they must not be enumerated. It might be possible to treat import rules that introduce new constants into a domain of abstract type. However, this case is not considered in the current implementation; it is left as a task for future work.

With these restrictions, the model checking approach for ASM works not only in principle, it works in practice too, as examples show (see for instance chapter 7.1 and 7.2). In most cases, however, we have to adapt the model at hand in order to fit it to the model checking process. The following section summarises our experiences and its consequences for model checking ASM. In Section 5.2, we give an insight into the transformation algorithm that transforms ASM models into a model checker language.

5.1   Modelling with Assumptions

Most ASM — on a certain level of abstraction — not only comprise the specification of state and state transitions, but have some additional *assumptions*. This is due to the fact that modelling on an abstract level provides concise and understandable models at the price of abstracting details that may influence the systems behaviour. Often assumptions on these abstracted details are necessary for proving correctness of the system. Figure 5.2 shows the layers of different abstraction levels, a modelling practice that is already introduced in Section 2.1.2. The abstracted information that is essential for the system's behaviour is kept in *assumptions* that accompany the structure and transition specification of an ASM (denoted in the figure as ASM_0, ASM_1, etc). These assumptions may concern system inherent details or the external world, i.e., the environment of the system.

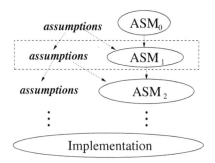

Figure 5.2: Different Layers of Abstraction

Assumptions related to system inherent behaviour may be implemented in some lower layer, at which the model is more concrete. We say the model is *refined*. Refinement is depicted in Figure 5.2 by means of dotted arrows that lead from one layer to the next. For the assumptions, the dotted lines indicate that through a refinement step some assumptions are absorbed by the ASM model on the lower level, others are kept as assumptions. Eventually the refinement leads ideally to an implementation that can be executed on a machine. The ground ASM model, the model at the topmost layer (ASM_0 in Figure 5.2), yields a basis for specifying the details of the assumptions explicitly later on. Generally, we can use the ASM language if we want to add the (otherwise lost) information to the ASM model.

Assumptions on the external environment of the system do not become part of the system specification. They are rather stated in a declarative way. This is common practice for the development of embedded systems that are not regarded to be robust against arbitrary — possibly faulty — environment behaviour. That is, we have to restrict the environmental behaviour the system is able to cope with. However, we cannot specify the entire environment explicitly. If we want to add this information about environment, the ASM language is not suitable. The operational semantics of ASM forces us to explicitly describe what happens, but a particular restriction to a certain kind of behaviour is hard to express. It is much more convenient to use a kind of logic for keeping this information.

We give some simple examples for abstracted information that is given within assumptions:

- In the ASM model of the bakery algorithm ([BGR94]) the ticket function T that maps to each new customer a new ticket is abstract in the ground model. It is specified as an external function. As an assumption over this function we need to assure that the number of a newly given ticket is greater than that of all other tickets that are already in the system. This way we guarantee a sensible order for serving the customers. This information is formalised by a logical formula. We assume that our external function T satisfies this logical property. In the more concrete ASM model, the function T is "implemented", i.e., it is explicitly modelled how T is updated during a run of the model. The external function becomes an "internal" one.

- Also within the ASM model of the bakery algorithm, we find that a fairness assumption is necessary for proving correctness of the algorithm ([BGR94]). At the moment there is no feature in the ASM language for expressing fairness assumptions. Instead the fair behaviour of processes is stated informally in the proof part.

- For embedded systems like the production cell ([Mea96] describes an ASM model of this case study), the specification of the environment is abstract and not part of the ordinary transition system. The behaviour of sensors is formalised by means of oracles or external functions. However, it is necessary to assume that the behaviour of the environment is "reasonable" in order to guarantee correctness of the ASM model. In [Mea96] most assumptions are given in terms of logical formulas that remain external also for the concrete ASM model.

- For the specification of protocols, we might abstract away the underlying communication model governing the transfer of messages, like in the ASM model of the FLASH cache coherence protocol in [Dur98]. But for the proofs we have to assume that the messages are transfered according to a particular strategy (e.g. the FIFO-strategy, so that the ordering of messages is preserved). At some lower level in the ASM hierarchy a proper message passing behaviour has to be specified that implements the assumption made on the order of transfered messages (cf. Section 7.1.3).

Obviously, only states, state transitions, and assumptions on the model together give the complete specification of the problem at hand. This view is sketched in Figure 5.2: the dashed box comprises all parts of a model at a particular level of abstraction. Our transformation algorithm for ASM models treats the structure and the transition rules of an ASM model. Additional information in the proof part or in the textual part, including logical formulas, is not considered. Thus, this information is lost after the transformation.

5.1.1   Model Checking without Meeting the Assumptions

Abstract behaviour is specified by oracles which are not influenced by the system itself but by some outside world. They introduce non-determinism. In ASM terms oracles are called external functions. External functions – in the case of finite domains – can simply be transformed into corresponding functions or variables that are not restricted, i.e., neither their initialisation nor their updating is specified. Since model checking is "input-less" it is a complete test over the whole state space. Any "loose" or non-restricted specification leads to a complete case distinction over all possibilities. The problems arising with this are not only a matter of the well-known state explosion problem, but also of getting useless counterexamples: If we fail to meet the additional assumptions on external functions, then the transformed model that is model checked, comprises behaviour that is excluded in the *complete* ordinary ASM model.

On the other hand, if we refine the model such that the assumptions are specified in terms of transition rules, it is essential to prove that the refinement is correct. Especially, we have to make sure that the refined model is not too restrictive. An environmental behaviour that has a high degree of freedom cannot be modelled properly in an operational way by using transition rules. Usually, it is too complex to be fully specified and only some of the actually allowed behaviour is given. Model checking a refined model that is too restrictive will potentially miss counterexamples that would arise from behaviour that is not specified and thus excluded.

The situation can be formalised as follows. Let $\mathcal{M}_{ASM}$ be the set of ASM models, $\mathcal{M}_{MC}$ be the set of models that are input to the model checker, and $\mathcal{RUN}$ be the set of runs of an ordinary temporal structure over branching time. The latter provides a semantics that both kinds of model have in common. To give a more precise notion of adding assumptions, we introduce mod and $\overline{mod}$ as two functions that yields the semantics of a model such that $mod : \mathcal{M}_{ASM} \rightarrow \mathcal{RUN}$ and $\overline{mod} : \mathcal{M}_{MC} \rightarrow \mathcal{RUN}$. Problems may arise in two cases:

$$mod(\mathcal{M}_{ASM}) \subset \overline{mod}(\mathcal{M}_{MC}) \tag{5.1}$$

$$mod(\mathcal{M}_{ASM}) \supset \overline{mod}(\mathcal{M}_{MC}) \tag{5.2}$$

In the first case, relation (5.1) expresses that the transformed MC model is strictly greater than the ASM model because of a loose specification of assumptions. The set $\overline{mod}(\mathcal{M}_{MC}) \setminus mod(\mathcal{M}_{ASM})$ may contain runs, that violate the property to be checked. One of these will produce a counterexample that is not a proper run of the ASM model. We can make no proposition on the model under investigation since only one counterexample will be given. We call it a *wrong counterexample*; it obstructs the overall debugging process.

In the second case, relation (5.2), we may fail to detect errors that occur only in those runs of the ASM model that are excluded for the MC model. The runs in $mod(\mathcal{M}_{ASM}) \setminus \overline{mod}(\mathcal{M}_{MC})$ will not be checked by the model checker.

We treat the problem of missing assumptions within the case studies that we examined (see Section 7.1 and 7.2). The solutions which we suggest depend on the model checking tool in use. In the first approach, using the model checker SMV, we model some assumptions explicitly by means of transition rules and conclude with some restrictions for embedded systems (c.f., chapter 6). For fairness assumptions the tool offers a special language construct.

In the second approach, using the MDG-package, we aim to include logical parts which formalise the assumptions into the model that is input to the model checker. This can be done by using linear temporal logic (LTL) for specifying the assumptions. As described in Section 3.4 on page 38, we can construct a tableau (which is a kind of transition system) from LTL formulas. The product automaton of the ASM model under investigation and the tableau of the LTL formula precisely specifies the ASM model plus the assumptions. The treatment of LTL with MDGs is already investigated in [Xu99]. For further discussion see Chapter 8.

In the next section, we introduce the algorithm that transforms ASM transition rules.

5.2   Transformation of Abstract State Machines

For the transformation, we assume that the ASM model is given in terms of domains, functions and transition rules. That is, either we use a correctly refined model that contains all additional assumptions, or the assumptions may be added later to the model that is input for the model checker.

The development of the transformation depends naturally on the choice of the tool we want to interface and its input language. For the sake of an efficient transformation, and a resulting model that is as simple as possible, we choose a tool among those model checkers that are based on transition systems (e.g., SMV [McM93], VIS [Gro96], SVE [F$^+$96]), rather than model checkers that are treating process algebras (e.g., FDR [For96] for CSP, the Concurrency Workbench [MS] for CCS, or CADP [FGM$^+$92] for LOTOS) or Petri nets for instance (e.g., the PEP-tool [GB96]). Our choice does not imply that using these model checkers of a different kind is generally impossible; only that the transformation would not be as straightforward and close as in our case since ASM are basically transition systems. Therefore, we relate the following considerations to the set of model checkers for transition systems among all others.

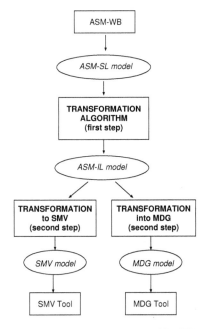

Figure 5.3: A generic Interface for ASM-WB

Since different model checkers (for transition systems) have different qualities, we are interested in having more than one interface to one particular tool. Therefore, we develop a more generic interface for the ASM-WB. Figure 5.3 shows the framework of algorithms that link the ASM-WB to two model checkers. The generic interface is realised by splitting the transformation into two major steps:

1. In the first step, the ASM model is transformed into a flat, simple transition system (in Figure 5.3 the ASM-IL model). The adjective "flat" indicates that there are no nested rule structures left, and "simple" means that all complex data structures are unfolded. This is a common representation for many analysis tools treating transitions systems and it is tool independent.

2. The second step for any interface depends on the tool. It provides a transformation from the simple transition system into the syntax of the input language of the chosen tool (in Figure 5.3 denoted as the SMV model and the MDG model). In most cases, this can be done easily.

We use the ASM-WB ([Cas99]) as a core tool for editing, parsing, type-checking and simulating models that are given in ASM-SL, a typed version of the ASM notation. In our work, we suggest the use of the SMV model checker ([McM93]) and the MDG-package ([CZS+97]) (cf. Figure 5.3). Obviously, this choice is not exhaustive. Within the workbench, a model is represented in an abstract syntax, called ASM-AST. The transformation algorithm transforms, in the first step, a model that is given in ASM-AST into the *intermediate language*, called ASM-IL. An ASM-IL model provides a general interface for transition-system-based tools. An interface from an ASM-IL representation to SMV for instance, which computes the second step of the transformation, compiles an ASM-IL model into the syntax of the SMV language.

The *decoupling* of the transformation into two steps provides the following advantages:

- The underlying schema for the transformation into a particular tool language can easily be changed by tailoring the second transformation step only. For this concern no insight into the ASM-WB and its internal representation (ASM-AST) is necessary but rather a mere understanding of the transformation from ASM-IL into the language of the considered tool is sufficient. This task is much less complex.

- The interface can be reused for other tools by replacing the transformation from ASM-IL to the target language and by replacing the mapping from the counterexample into the simulator of the ASM-WB.

The disadvantage of this generic interface is given by the fact that with applying the first step of the transformation, we lose all information about the model structure. The ASM-IL format provides no means for expressing modularisation or hierarchical structure. For the approaches we use so far, however, we cannot benefit from structural information anyway. This claim is justified further in Chapter 6.

A similar approach for the use of a generic interface language can be found in many tool frameworks. A very high-level intermediate language is given by SAL ("the common intermediate language", introduced in [BGL+00] and [Sha00]), for instance. It is a formal language for which interfaces to various tools and algorithms are already available. In contrast to our work, the focus for the development of SAL was more tool-oriented: several tools were available and a common language was built on top of them. A transformation of the ASM language into SAL would be interesting in order to exploit the wide range of tool-interfaces. However, the transformation would not be as close as the transformation to the ASM-specific interface language ASM-IL.

A brief history of the development. In our first approach, we use the model checker SMV for model checking ASM. In [Win97] a first transformation schema from ASM into the SMV language is introduced. This schema is later used for developing the interface from the ASM-WB to SMV. The results are published as joint work with Giuseppe Del Castillo in [CW00]. [Win00] suggests, among other results, optimisations to the original algorithm in order to make the transformation feasible for large ASM models. For the use of MDGs, we extend the interface further. This extension is introduced in [Win01].

In the following subsection, we give a definition of the mapping from ASM models into the intermediate language. We point out in which way the transformation is optimised in order to be feasible for large ASM models. The section ends with the proof of correctness of the transformation.

5.2.1   The Transformation Algorithm

Simple transition systems treated by a model checker are specified in terms of a set of states $\mathcal{S}$ and a transition relation $\mathcal{R} : \mathcal{S} \to \mathcal{S}$ defined over states. A state is defined in terms of state variables s_i, $1 \leq i \leq n$. For each state variable s_i, we use the primed variable s_i' to denote the state variable in the next state. $\mathcal{R}$ defines a set of pairs $(pre(\bar{s}), post(\bar{s}'))$ with $pre(\bar{s})$ as a pre-state of a transition given in terms of (unprimed) state variables and $post(\bar{s}')$ as a post-state of a transition given in terms of next state variables $\bar{s}'$.

Many model checker languages describe the transition relation with respect to changes on each state variable individually (e.g., the SMV language, [McM93]), i.e., we get a set of pairs of the form $(pre(\bar{s}), post(s_1, \ldots, s_{i-1}, s_i', s_{i+1}, \ldots, s_n))$. The post-state depends on the state variables that remain unchanged, and a single next state variable that is changed, s_i'.

In ASM, the same information can be given as simple guarded updates of the form

$$\text{if } g \text{ then } loc_i := val. \tag{5.3}$$

Following Section 2, the semantics of this guarded update is given with respect to an expanded state $B = (A, \zeta)$, $\Delta_B(loc_i := val)$ if $B \models g$. That is, the guard g specifies the set of states in which the transition fires. This is denoted above as the pre-state of the transition, $pre(\overline{loc})$, where $\overline{loc} = (loc_1, \ldots, loc_n)$ is the tuple of locations of the ASM.

The semantics of the update $\Delta_B(loc_i := val)$ is given as a singleton $\{(loc_i, val)\}$. Implicitly, each other location keeps its value. This specifies the set of post-states, i.e., $post(loc_1, \ldots, loc_{i-1}, loc_i', loc_{i+1}, \ldots, loc_n)$.

The identification of the notations, pre- and post-states and ASM updates, requires the following two conditions:

1. We can identify the notion of locations with state variables, i.e., $loc_i \equiv s_i$. Following the definition (2.11), a *location* is a pair $loc = (f, \bar{a})$, where $\bar{a}$ is a tuple of elements of S^A. That is, a location is a function applied to particular parameter values. Thus, each location can be identified with a state variable. A renaming function maps each location pair $(f, \bar{a})$ to a unique variable name.

2. The guard is a simple boolean term that depends only on unprimed locations, i.e., $g(\overline{loc})$.

Any ASM (with some restrictions as shown below) can be transformed into a set of guarded updates of form of (5.3) that satisfies these requirements. We use simple guarded updates as in (5.3) as our intermediate language (ASM-IL) for representing ASM models. A shorter representation is given through the triple $(loc_i, guard, val)$. Since each location can be affected by different transition rules, each loc_i is, in fact, attached with a *set* of guarded updates, $(loc_i, \{guard_j, val_j \mid 1 \leq j \leq n\})$. If the set of locations of an ASM model is given as $\{loc_1, \ldots, loc_m\}$ then the behaviour of the model can be represented as a set of guarded updates over all locations $\{(loc_i, \{guard_j, val_j \mid 1 \leq j \leq n\}) \mid 1 \leq i \leq m\}$. For the transformation of ASM transition rules into the ASM-IL notation, we have to *unfold the dynamic functions* into locations and *simplify the rule structure*.

Unfolding dynamic functions. The transformation into ASM-IL requires unfolding of all dynamic and external functions occurring in the transition rules of a model into locations.

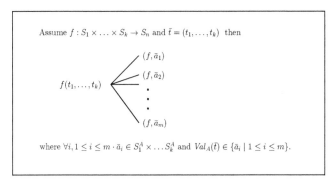

Figure 5.4: Unfolding dynamic functions

Usually, the left hand side (LHS) of an update in an ASM transition rule is given as a term $f(t_1, \ldots, t_k)$, where f is a dynamic function symbol. The parameter terms t_i can be dynamic or external functions as well, i.e., their evaluation is state dependent. For generating a model for the model checker the locations have to be evaluated (and declared) before rather than during the state exploration of the checking process. Therefore, we have to unfold the terms into every possible value that might occur during a run (see Figure 5.4). Recall that $f : S_1 \times \ldots \times S_k \to S_n$ specifies the type of the function symbol f. S_i^A is a universe or domain (of sort S_i), i.e., a set of entities, in a particular state A. $Val_A(\bar{t})$ is the evaluation of a term $\bar{t}$ in state A.

It is easy to see that the use of dynamic functions of large arity k increases the effort of unfolding exponentially. The number of locations unfolded from a function f of type $S_1 \times \ldots \times S_k \to S_n$ is given with respect to a state A by the product $card(S_1^A) \cdot \ldots \cdot card(S_k^A)$, where function $card$ gives the cardinality of the parameter domains.[1] Lists of parameters for dynamic functions should be used with care when modelling with ASM. Moreover, it is obvious that the domains S_i^A in Figure 5.4 have to be finite. Otherwise the number of locations would be infinite theoretically, and in practice the unfolding would not terminate.

The actual application of unfolding non-static terms is spread over two functions, the *term simplification* $[\![.]\!]$ and the *rule unfolding* $\mathcal{E}$ which we introduce below. The term simplification is inductively defined over rules and terms.

---

[1]Note that for any universe S_i^A, $card(S_i^A)$ is constant for all states A since we exclude expanding universes. Universes have to be fixed.

71

Simplification function

For transforming an ASM model into ASM-IL format, we have to map each transition rule into a set of simple updates. That is, nested rules have to be flattened and other rule constructs have to be mapped into guarded updates. We describe this mapping as a function $[\![.]\!]_\varsigma$, called *simplification function*, that maps ASM-rules into the simple guarded update notation of ASM-IL and also simplifies unfolded ASM-IL rules further (see rule unfolding function $\mathcal{E}$ below). It is defined in terms of a variable assignment $\varsigma : V \to S^A$, which maps each free variable to a value (c.f., definition 2.6). This definition coincides with the elaboration in [Cas00]. Firstly, we give a definition of the simplification function $[\![.]\!]_\varsigma$ for terms and for transition rules. At the end of this section, we prove for each rule R that the simplification function preserves the semantics of R.

Simplifying terms.   The simplification mapping $[\![.]\!]_\varsigma$ maps all terms that occur in the ASM rules, for instance, the LHS and the right hand side (RHS) of an update rule, or guards in conditional or `do-for-all` rules. Mapping terms provides the inductive base of the simplification function and ensures that the application of $[\![.]\!]_\varsigma$ to any rule over finite domains terminates.

We distinguish three base terms: constant *values*, *locations* that can be identified with state variables, and *variables* that occur free. The simplification of a term terminates if the result consists of only these base terms. All other terms have to be simplified further. This is captured in the following points.

- Values (or constants) of any sort $a \in S_i^A$ (they are constant in any state A) are not affected by the simplification[2]. (For convenience from now on we denote values with a.)
$$[\![a]\!]_\varsigma = a$$

- Locations $loc = (f, \overline{a})$ are not affected by the simplification either[3]:
$$[\![\, loc \,]\!]_\varsigma = loc$$

- Variables are mapped to values by the variable assignment ς if they are in its domain.
$$[\![\, v \,]\!]_\varsigma = \begin{cases} a = \varsigma(v) & \text{if } v \in \text{dom}(\varsigma) \\ v & \text{otherwise} \end{cases}$$

- For function applications we distinguish two cases: either all parameters of the function are values or some of the parameters have to be simplified further.

   a.) Function applications which comprise only values as parameters are simplified to values if the function symbol is static and can be evaluated independently of the state. If the function symbol is dynamic or external, i.e., its evaluation is state-dependent, the term is mapped to a location $(f, \overline{a})$.

   $[\![t_i]\!]_\varsigma = a_i$ for each $i \in \{1, \dots, n\}$ $\Rightarrow$

   $$[\![f(t_1, \dots, t_n)]\!]_\varsigma = \begin{cases} a = f^A(a_1, \dots, a_n) & \text{if } f \text{ is a static function name} \\ loc = (f, (a_1, \dots, a_n)) & \text{if } f \text{ is a dynamic/external} \\ & \text{function name} \end{cases}$$

---

[2] Note that constants do not occur in an ASM but in ASM-IL. The simplification function can be applied to a in the second simplification step (after unfolding a rule by $\mathcal{E}$).

[3] For locations holds the same, they do not occur in ASM but in ASM-IL. A second application of $[\![.]\!]_\varsigma$ is possible after rule unfolding.

where f^A denotes the interpretation of a static function symbol f in any state A.

b.) For terms which comprise non-values, i.e., locations or not yet evaluated functions, the term simplification is applied to the parameters.

$$[\![t_i]\!]_\varsigma = loc \text{ or } [\![t_i]\!]_\varsigma = f'(\bar{t}) \text{ for some } i \in \{1,\ldots,n\} \Rightarrow$$
$$[\![f(t_1,\ldots,t_n)]\!]_\varsigma = f([\![t_1]\!]_\varsigma,\ldots,[\![t_n]\!]_\varsigma)$$

- First order terms $(\exists v : g(v))s(v)$ and $(\forall v : g(v))s(v)$ can be simplified if we assume that the guard of the first order term $g(v)$ has a finite range, i.e., $Range_A(v : g(v)) = \{a \in S_i^A \mid A \models g(a)\} = \{a_1,\ldots,a_n\}$ is a finite set. In this case, the variable assignment ς is refined to a mapping that maps the head variable v to values of $Range_A(v : g(v))$, i.e., $\varsigma[v \mapsto a_i]$ for $1 \leq i \leq n$. The body $s(v)$ is further simplified with respect to this refined variable assignment. Existential quantification naturally leads to a disjunction over all assignments of v, and universal quantification to a conjunction.

$$[\![((\exists v : g(v)) \; s(v))]\!]_\varsigma = [\![s(v)]\!]_{\varsigma[v \mapsto a_1]} \vee \ldots \vee [\![s(v)]\!]_{\varsigma[v \mapsto a_n]}$$

$$[\![((\forall v : g(v)) \; s(v))]\!]_\varsigma = [\![s(v)]\!]_{\varsigma[v \mapsto a_1]} \wedge \ldots \wedge [\![s(v)]\!]_{\varsigma[v \mapsto a_n]}$$

Simplifying transition rules   Once the inductive base of the simplification function $[\![.]\!]_\varsigma$ is defined, we can continue with the simplification of transition rules.

1. The skip rule is not effected by the simplification:

$$[\![\text{skip}]\!]_\varsigma = \text{skip}$$

2. An update rule is simplified by the unfolding of the LHS as described above. The RHS might depend on dynamic functions as well and is therefore treated in the same way. As a default guard we introduce *true* to indicate that the precondition is not restricted:

$$[\![f(\bar{t}) := t]\!]_\varsigma = \text{if } \textit{true} \text{ then } [\![f(\bar{t})]\!]_\varsigma := [\![t]\!]_\varsigma$$

3. A block-rule is simplified by means of simplifying each of its sub-rules:

$$[\![\text{block } R_1 \; \ldots \; R_n \text{ endblock}]\!]_\varsigma = [\![R_1]\!]_\varsigma \; \ldots \; [\![R_2]\!]_\varsigma$$

4. Simple conditional rules give a collection of guarded updates. The update of each location in the then-case or in the else-case is considered separately with respect to the guard and its negation respectively:

$$
\begin{aligned}
&[\![\text{if } g \\
&\text{then } f_1(\overline{t_1}) := v_1 \\
&\qquad \ldots \\
&\qquad f_n(\overline{t_n}) := v_n \\
&\text{else } g_1(\overline{s_1}) := w_1 \\
&\qquad \ldots \\
&\qquad g_m(\overline{s_m}) := w_m \\
&\text{endif}]\!]_\varsigma
\end{aligned}
=
\begin{cases}
\text{if } [\![g]\!]_\varsigma \text{ then } [\![f_1(\overline{t_1})]\!]_\varsigma := [\![v_1]\!]_\varsigma \\
\qquad \ldots \\
\text{if } [\![g]\!]_\varsigma \text{ then } [\![f_n(\overline{t_n})]\!]_\varsigma := [\![v_n]\!]_\varsigma \\
\text{if } \neg[\![g]\!]_\varsigma \text{ then } [\![g_1(\overline{s_1})]\!]_\varsigma := [\![w_1]\!]_\varsigma \\
\qquad \ldots \\
\text{if } \neg[\![g]\!]_\varsigma \text{ then } [\![g_m(\overline{s_m})]\!]_\varsigma := [\![w_m]\!]_\varsigma
\end{cases}
$$

For nested conditional rules we have to collect, for each internal update, all influencing guards or their corresponding negation:

$$
\llbracket \text{if } g_1 \text{ then} \atop \quad \begin{array}{l} \text{if } g_2 \text{ then } f_1(\overline{t_1}) := v_1 \\ \ \dots \text{ endif} \\ \text{else} \\ \quad \text{if } g_3 \text{ then } h_1(\overline{p_1}) := r_1 \\ \quad \dots \text{ endif} \\ \text{endif} \rrbracket_\varsigma \end{array}
\ = \
\begin{cases} \text{if } \llbracket g_1 \rrbracket_\varsigma \wedge \llbracket g_2 \rrbracket_\varsigma \text{ then } \llbracket f_1(\overline{t_1}) \rrbracket_\varsigma := \llbracket v_1 \rrbracket_\varsigma \\ \qquad \dots \\ \text{if } \neg \llbracket g_1 \rrbracket_\varsigma \wedge \llbracket g_3 \rrbracket_\varsigma \text{ then } \llbracket h_1(\overline{p_1}) \rrbracket_\varsigma := \llbracket r_1 \rrbracket_\varsigma \\ \qquad \dots \end{cases}
$$

Note that the guards have to be simplified as well since in the intermediate representation they should depend only on locations and not on n-ary functions.

This schema is inductively captured for all possible inner rules as

$$
\llbracket \text{if } g \text{ then } R_1 \text{ else } R_2 \rrbracket_\varsigma \ = \ \begin{cases} \text{if } \llbracket g \rrbracket_\varsigma \text{ then } \llbracket R_1 \rrbracket_\varsigma \\ \text{if } \neg \llbracket g \rrbracket_\varsigma \text{ then } \llbracket R_2 \rrbracket_\varsigma \end{cases}
$$

5. `do-for-all` rules can be mapped into a collection of rules that are instantiated for each possible value of the head variable. For a rule as given below, the value of the head variable v is chosen among values in its range $Range_B(v : g(v)) = \{a \in S_i^A \mid B \models g(a)\}$. We assume that the sort S_i^A is a finite set $\{a_1, \dots, a_m\}$. The guard $g(v)$ in the head of the rule is used – after instantiating the parameter v with its value – as a guard for the simple guarded update.

$$
\llbracket \begin{array}{l} \text{do forall } v \ : \ g(v) \\ \quad R_0(v) \\ \text{enddo} \rrbracket_\varsigma \end{array}
\ = \
\begin{cases} \text{if } \llbracket g(a_1) \rrbracket_\varsigma \text{ then } \llbracket R_0(a_1) \rrbracket_\varsigma \\ \qquad \dots \\ \text{if } \llbracket g(a_m) \rrbracket_\varsigma \text{ then } \llbracket R_0(a_m) \rrbracket_\varsigma \end{cases}
$$

6. A `choose`-rule is not treated by our transformation but it can be "simulated" by means of an external function ext that is element of $Range_B(v : g(v))$ (see Definition 2.9). This external function ext is non-deterministically evaluated to the parameter value to be chosen:

$$
\begin{array}{l} \text{choose } v \ : \ g(v) \\ \quad R_0(v) \\ \text{endchoose} \end{array}
\quad \Leftrightarrow \quad \text{if } \llbracket g(ext) \rrbracket_\varsigma \text{ then } \llbracket R_0(ext) \rrbracket_\varsigma
$$

where the range of the head variable v is $Range_B(v : g(v)) = \{a \in S_i^A \mid B \models g(a)\}$ and ext is an external nullary function of the same type as v, i.e., S_i^A.

7. `import`-rules are not treated by the transformation since they extend the domains of the model. Model with non-fixed state space can not be treated by the model checkers we investigated since this requires the declaration of infinite domains in general.

Rule unfolding

Simplified rules and terms may still contain functional terms with locations as parameters, e.g., $f(\dots, loc_j, \dots)$. These terms have to be *unfolded* such that each location in the parameter list is substituted with a possible value of its domain (c.f., Figure 5.4). For each of these substitutions, we generate a new instance of the rule in which the location occurs as a parameter. This procedure is given as the unfolding function which is described below.

Unfolding of rules. When we unfold transition rules we replace all locations that appear as parameters of functions with their possible evaluation. Locations that occur on the LHS of an update are left unchanged. We define the function $\mathcal{E}$ that maps rules to unfolded rules.

- A rule that contains only simple updates of locations, i.e., no location occurs on the RHS of an update, remains unchanged by the rule unfolding function.

$$\mathcal{E}(R) = R \quad \text{if } R = (loc_1 := a_1 \ \dots \ loc_n := a_n)$$

- A rule where loc is the first location occurring in R but not as an LHS of an update rule is unfolded into n instances when the range of the location loc is $\{a_1, \dots, a_n\}$. The generated rules instances are simplified further by means of applying function $[\![.]\!]_\varsigma$.

$$\mathcal{E}(R(loc)) = \left\{ \begin{array}{l} \texttt{if } loc = a_1 \texttt{ then } \mathcal{E}([\![R[loc/a_1]]\!]_\varsigma) \\ \dots \\ \texttt{if } loc = a_n \texttt{ then } \mathcal{E}([\![R[loc/a_n]]\!]_\varsigma) \end{array} \right.$$

The notation $R[loc/a_i]$ denotes an instance of rule R in which every occurrence of loc is substituted by value a_i.

For every possible value a_i of a location, a new rule instance has to be generated which is guarded by the equation $loc = a_i$. This causes alot of code to be duplicated and the resulting model can grow very big. This problem is address below.

Optimisation. The unfolding of rules may generate of lot of ASM-IL code due to the fact of duplicating rules while instantiating locations. Since unfolding and simplification are interleaved in our implementation, a large part of the ASM-IL rules can be simplified. In particular, the simplification of guards reduces the code significantly. (Note that guards are boolean terms that can be evaluated.) For instance, the (unfolded) guard $(time = 10) \wedge (time > 20)$, of course, evaluates to *false*.

More often occurs the case in which an equation over a simple location is unfolded. For example let *mode* be a simple location that may evaluate to *stop*, *run*, and *wait*. A guard of the form $(mode = stop)$ is firstly unfolded into three instances of the guard: $(mode = stop) \wedge (stop = stop)$, $(mode = run) \wedge (run = stop)$, and $(mode = wait) \wedge (wait = stop)$. The second and the third instance are simplified to *false* and the resulting instance of the guard is $(mode = stop) \wedge true$, which is obviously correct.

Still the effort of unfolding and simplifying is significant. A first implementation without any optimisation led to a bottleneck that obstructed, in some cases, the whole model checking approach.

The crucial point for a successful optimisation of the algorithm is that the functional terms to be unfolded should be chosen with care. Unnecessary unfolding has to be avoided. For instance (this probably caused the major problem; compare the example above), a simple equation in a guard $loc = a$ is internally (i.e., in ASM-AST) represented as a function application with a location as parameter, i.e., $=(loc, a_i)$. Thus, this term is unfolded and n rule instantiations are generated if loc ranges between $a_1, \dots, a_n$ (see Figure 5.5). All unfolded guard terms are simplified again to *false* except the one that evaluates to *true* when the simplification is applied. The result will be one rule instance only. However, this unnecessary unfolding and simplifying is very complex, especially since these kind of equations occur quite often in ASM rules. Therefore unfolding of equational terms, for instance, has to be avoided.

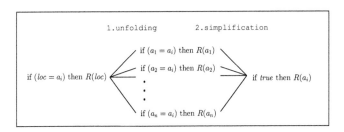

Figure 5.5: Unnecessary unfolding of equations

Based on this preliminary understanding, we conclude with the following more general solution. All symbolic model checkers are able to treat at least simple boolean operations for equality, conjunction and disjunction since these can be represented by decision diagrams. In the input language of SMV model checker, also simple arithmetic operations are defined (see [McM93]). We define the set of operations that are treated by the model checker as *primary operations*. Terms that denote these operations must not be unfolded even if they contain locations. They are simply mapped into the proper model checker syntax later on (i.e., in the second step of the transformation). That is, for any term $f(t_1, \ldots, t_n)$ where some of the parameters t_i are locations we test if f is a function symbol that denotes a primary operation. If it is, then we keep it, otherwise we unfold the location.

Multi-Agent ASM   We treat multi-agent ASM, which models concurrent or distributed systems, by means of an interleaving semantics.

The basic idea of multi-agent ASM is that the system consists of several *agents* identified with the elements of a finite set $AGENT$ (which are actually a sort of "agent identifiers"). Each agent $a \in AGENT$ executes its own parameterised program $prog(a)$. As a general parameter, we use a special nullary function $Self : AGENT$.

To map a multi-agent ASM model with agents $a_i \in AGENT$ into ASM-IL notation, we consider parameter $Self$ as an external function which determines the active agent. We assume that there is one program P, shared by all agents, possibly performing different actions for different agents. The FLASH model presented in Section 7.1 is an example of this style of modelling in which all agents execute exactly the same program, but on different data.

Due to the external function $Self$, which is treated as a location that has to be unfolded, a multi-agent model is mapped into a collection of ASM-IL rules that are guarded by the evaluation of $Self$, e.g.,:

> **if** $Self = a_1$ **then** $prog(a_1)$
>
> $\ldots$
>
> **if** $Self = a_n$ **then** $prog(a_n)$

where $\{a_1, \ldots, a_n\}$ are the agents and $prog(a_i)$ is the rule to be executed by agent a_i, i.e., the "program" of a_i. Since $Self$ is an external function its value is chosen non-deterministically and denotes the agent that is currently active. This results in a model with an interleaving semantics which allows any order for the sequence of active agents.

Based on the representation of models in ASM-IL which is provided by performing the first step of the transformation, the second step of the transformation generates a model in the proper syntax of the chosen model checker. The second step is thus tool dependent and is introduced in the corresponding Chapters 6 and 8 for the interface to SMV and the MDG-package, respectively.

Correctness of the simplification function

Simplification and unfolding as given above transform an ASM model into an ASM-IL model. This transformation is not a mere syntactical mapping since terms are partly evaluated: static terms are evaluated by $[\![.]\!]_\zeta$, and locations are unfolded to values by $\mathcal{E}$. Due to this fact, the semantics of ASM (see the definition in Chapter 2) is slightly different to the semantics of ASM-IL. We define Δ'_B as deterministic semantics and $\mathcal{N}'_B$ as non-deterministic semantics for ASM-IL as follows:

Definition 5.22:   Semantics for ASM-IL

For any ASM-IL rule $R \equiv loc := val$, where loc is a location and val is a value, the deterministic semantics is given as

$$\Delta'_B(R) = \{(loc, val)\}$$

and the non-deterministic semantics is given as

$$\mathcal{N}'_B(R) = \{\Delta'_B(R)\}.$$

For any other ASM-IL rule R' $\Delta'_B(R') = \Delta_B(R')$; $\mathcal{N}'_B(R')$ is equal to the definition of $\mathcal{N}_B(R')$ with every occurrence of $\Delta_B(.)$ replaced by $\Delta'_B(.)$. Note that $\mathcal{N}'_B$ can be applied to several rules:

$$\mathcal{N}'_B(R_1, \ldots, R_n) = \{setTuple(t) \mid t \in \mathcal{N}'_B(R_1) \times \ldots \times \mathcal{N}'_B(R_n)\}$$

where $setTuple(s_1, \ldots, s_n) = s_1 \cup \ldots \cup s_n$.

Following the definition above, the only difference between these semantics is given by the evaluation of terms in update rules. Terms in ASM-IL are already evaluated. Apart from that, the semantics is the same for both languages.

We now prove the correctness of the simplification of terms and of transition rules. Note that $[\![.]\!]_\zeta$ applied to a constant a or a location loc (which are ASM-IL terms) has no effect.

Terms.   When applied to static functions or variables, the simplification function $[\![.]\!]_\zeta$ is identical to the evaluation function Val_B on terms (see Definition 2.8). Both functions are defined with respect to ζ, which evaluates variables, and both functions map static functions to their value. If the term is an application of a dynamic or external function then the simplification does not evaluate this term but maps it to a location.

The difference between the functions Val_B and $[\![.]\!]_\zeta$ is that $[\![.]\!]_\zeta$ maps terms independently of a state and Val_B maps terms with respect to a state. In fact, the term evaluation Val_B is defined with respect to an expanded state B, $B = (A, \zeta)$. Simplification $[\![.]\!]_\zeta$ maps terms whose evaluation depends on a state (i.e., applications of dynamic or eternal functions) to locations (see Definition 2.11). We give the proof of correctness of the simplification function $[\![.]\!]_\zeta$ in two steps:

a.) $[\![.]\!]_\zeta$ transforms, where possible, a term into $Val_B(t)$; for this case we distinguish the following:

- a nullary static function f is mapped to its interpretation, i.e., $[\![f]\!]_\zeta = f^A = Val_B(f)$;

- a variable v is mapped to its variable assignment, i.e., $[\![v]\!]_\zeta = \zeta(v) = Val_B(v)$;

- if f is a static function name and $\bar{a}$ is a tuple of values then $f(\bar{a})$ is simplified to $f^A(\bar{a})$ which is equal to $Val_B(f(\bar{a}))$;

- if f is a static function name and t_i are static terms (for $1 \leq i \leq n$) then $f(t_1, \ldots, t_n)$ is simplified to $f^A([\![t_1]\!]_\zeta, \ldots, [\![t_n]\!]_\zeta)$ which is equal to $f^A(Val(t_1), \ldots, Val_B(t_n)) = Val_B(f(t1, \ldots, t_n))$ by Definition 2.8.

b.) $[\![.]\!]_\zeta$ transforms dynamic/external terms t that cannot be evaluated independently of the state into locations (i.e., state variables) i.e., $[\![t]\!]_\zeta = loc_t$. We prove that this location has the value $Val_B(t)$ at any state A, where $B = (A, \zeta)$. Assume in the following that f is a dynamic or external function symbol, $B = (A, \zeta)$, $\bar{a} = (a_1, \ldots, a_n)$, and $loc_i = (f, (\bar{a}))$:

- If f is a nullary function, it is mapped to its location, i.e., $[\![f]\!]_\zeta = loc_f$. According to Definitions (2.11) and (2.8), this location has the value $f^A = Val_B(f)$ in any state A.

- If f is a n-ary function and $\bar{a}$ a tuple of values then $f(\bar{a})$ is simplified to the location $loc_i = (f, (\bar{a}))$. According to Definitions (2.11) and (2.8), loc_i has the value $f^A(\bar{a}) = Val_B(f(\bar{a}))$.

- It remains to prove that the semantics is also preserved for functions with dynamic parameters that are given in terms of dynamic parameters, i.e., if we have a nested application of dynamic functions. This can be done by induction over the depth n of nested dynamic function application in the term.

  Assume g_i, for $1 \leq i \leq n$, are dynamic functions. Without loss of generality, let all g_i be unary functions. Furthermore, we assume that the nesting of dynamic function application is finite. That is, for any term $g_{n+1}(g_n(\ldots(g_1(g_0(\bar{t})))\ldots))$, g_0 is a dynamic or static function that is applied to only values.

  Induction base: We proved already that $[\![g_0(\bar{t})]\!]_\zeta$ equals to $Val_B(g_0(\bar{t}))$.

  Induction step: Assume $t = g_n(\ldots(g_0(\bar{t}))\ldots)$ and $[\![t]\!]_\zeta$ is simplified into a location loc_{g_n}. We have to prove that $[\![g_{n+1}(t)]\!]_\zeta = Val_B(g(t))$: According to the definition of $[\![.]\!]_\zeta$, $[\![g_{n+1}(t)]\!]_\zeta = g_{n+1}([\![t]\!]_\zeta)$. With the induction hypothesis we get $g_{n+1}(loc_{g_n})$.

  The location loc_{g_n} is unfolded by the rule unfolding $\mathcal{E}$ to any possible value since it is not an LHS of an update. Thus, $g_{n+1}(loc_{g_n})$ is simplified and unfolded into a set of locations; for each value of loc_{g_n} we get a new location. Analogously to the proof given above, we can state that this set of locations is equal to the locations that are defined by $g_{n+1}(t)$. Their state-dependent interpretation is preserved.

c.) First order terms of the form $(\exists v : g(v))\, s(v)$ or $(\forall v : g(v))\, s(v)$ are simplified to their value. Assume $t = (\exists v : g(v))s(v)$ and $Range_B(v : g(v)) = \{a_1, \ldots, a_n\}$ (note that the range has to be finite for applying the simplification function, see the definition of $[\![.]\!]_\varsigma$). Then

$$[\![t]\!]_\varsigma$$
(according to the definition of $[\![.]\!]_\varsigma$)
$$= \quad [\![s(v)]\!]_{\varsigma[v \mapsto a_1]} \lor \ldots \lor [\![s(v)]\!]_{\varsigma[v \mapsto a_n]}$$
(according to the correctness proof above and Definition (2.7))
$$= \quad Val_{B(v \mapsto a_1)}(s(v)) \lor \ldots \lor Val_{B(v \mapsto a_n)}(s(v))$$
$$= \quad true, \text{ if } Val_{B(v \mapsto a_1)}(s(v)) \text{ for some } a \in \{a_1, \ldots, a_n\}$$
(according to first order term evaluation, Def. (2.10))
$$= \quad Val_B(t).$$

Similarly, it follows that $[\![(\forall v : g(v))\ s(v)]\!]_\varsigma = Val_B((\forall v : g(v))\ s(v))$.

Transition Rules. In ASM, the semantics of deterministic transition rules is defined through the operator Δ_B and that of non-deterministic transition rules through the operator $\mathcal{N}_B$ (see Chapter 2). Analogously, we defined the semantics for ASM-IL rules (see Definition (5.22)), Δ'_B and $\mathcal{N}'_B$. Since the transformation treats **choose-rules** as well, we have to use the semantics for non-deterministic transition rules to prove the correctness of the transformation of rules. We have to show that for any rule R, $\mathcal{N}_B(R) = \mathcal{N}'_B([\![R]\!]_\varsigma)$.

1. **skip** rules:

   Following the definition of $[\![.]\!]_\varsigma$, the skip rule is not affected by the simplification:

   $$\mathcal{N}'_B([\![\texttt{skip}]\!]_\varsigma) = \mathcal{N}'_B(\texttt{skip}) = \{\{\}\} = \mathcal{N}_B(\texttt{skip})$$

2. **update** rules:
   Assume $\bar{t} = (t_1, \ldots, t_n)$, $\bar{a} = (a_1, \ldots, a_n)$ and $Val_B(t_i) = a_i$ for every i, $1 \leq i \leq n$.

   $$\mathcal{N}'_B([\![f(\bar{t}) := t]\!]_\varsigma)$$
   $= \quad \mathcal{N}'_B(\texttt{if } true \texttt{ then } [\![f(\bar{t})]\!]_\varsigma := [\![t]\!]_\varsigma) \quad$ (according to the definition of $[\![.]\!]_\varsigma$)
   $= \quad \mathcal{N}'_B([\![f(\bar{t})]\!]_\varsigma := [\![t]\!]_\varsigma) \quad$ (for any state A since $A \models true$,
   $\qquad$ i.e., the guard is always satisfied)
   $= \quad \mathcal{N}'_B(\ (f, (\bar{a})) := Val_B(t)\ ) \quad$ (according to the definition of $[\![.]\!]_\varsigma$
   $\qquad$ and the proof of correctness above)
   $= \quad \{\Delta'_B((f, (\bar{a})) := Val_B(t))\} \quad$ (according to the definition of $\mathcal{N}'_B$)
   $= \quad \{\{((f, (\bar{a})), Val_B(t))\}\} \quad$ (according to the definition of Δ'_B)
   $= \quad \{\Delta_B(f(\bar{t}) := t)\} \quad$ (according to the definition of Δ_B)
   $= \quad \mathcal{N}_B(f(\bar{t}) := t) \quad$ (according to the definition of $\mathcal{N}_B$)

3. **conditional** rules:

   $$\mathcal{N}'_B([\![\texttt{if } g \texttt{ then } R_1 \texttt{ else } R_2]\!]_\varsigma)$$
   $= \quad \{setTuple(t) \mid t \in \mathcal{N}'_B(\texttt{if } [\![g]\!]_\varsigma \texttt{ then } [\![R_1]\!]_\varsigma) \times \mathcal{N}'_B(\texttt{if } \neg[\![g]\!]_\varsigma \texttt{ then } [\![R_2]\!]_\varsigma)\}$
   $\qquad$ (according to definition of $[\![.]\!]_\varsigma$ and Definition (5.22))

   - If $B \models [\![g]\!]_\varsigma$, this evaluates to

   $\{setTuple(t) \mid t \in \mathcal{N}'_B([\![R_1]\!]_\varsigma) \times \mathcal{N}'_B([\![skip]\!]_\varsigma)\}$
   $= \quad \{setTuple(t) \mid t \in \mathcal{N}'_B([\![R_1]\!]_\varsigma) \times \{\{\}\}\}$
   $= \quad \mathcal{N}'_B([\![R_1]\!]_\varsigma) \quad$ (according to Definition (5.22))

- If $B \models \neg[\![g]\!]_\varsigma$, this evaluates to

$$\{setTuple(t) \mid t \in \mathcal{N}'_B([\![R_2]\!]_\varsigma) \times \mathcal{N}'_B([\![skip]\!]_\varsigma)\}$$
$$= \{setTuple(t) \mid t \in \mathcal{N}'_B([\![R_2]\!]_\varsigma) \times \{\{\}\}\}$$
$$= \mathcal{N}'_B([\![R_2]\!]_\varsigma) \qquad \text{(according to Definition (5.22))}$$

If we assume that $\mathcal{N}'_B([\![R_i]\!]_\varsigma) = \mathcal{N}_B(R_i)$ for $i \in \{1, 2\}$, it remains to prove that in any expanded state B

$$B \models [\![g]\!]_\varsigma \;\Leftrightarrow\; B \models g.$$

With the proof of the correctness of term simplification given above, we get $[\![g]\!]_\varsigma = Val_B(g)$ for any boolean term or any first order term g. It follows that g and $[\![g]\!]_\varsigma$ are satisfied in the same states and thus

$$\mathcal{N}'_B([\![\texttt{if } g \texttt{ then } R_1 \texttt{ else } R_2]\!]_\varsigma) = \mathcal{N}_B(\texttt{if } g \texttt{ then } R_1 \texttt{ else } R_2).$$

4. **do-forall** rules:

Assume that $a_i \in Range_B(v : g(v))$, where $1 \leq i \leq n$.
$$\mathcal{N}'_B([\![\texttt{do forall } v \;:\; g(v) \;\; R_0(v) \texttt{ enddo}]\!]_\varsigma)$$

$$= \{setTuple(t) \mid t \in \mathcal{N}'_B(\texttt{if } [\![g(a_1)]\!]_\varsigma \texttt{ then } [\![R_0(a_1)]\!]_\varsigma)$$
$$\times \ldots \times \mathcal{N}'_B(\texttt{if } [\![g(a_n)]\!]_\varsigma \texttt{ then } [\![R_0(a_n)]\!]_\varsigma)\}$$
(according to the definition of $[\![.]\!]_\varsigma$ and Definition (5.22))

Assuming without loss of generality that $Range_B(v : g(v)) = \{a_1, \ldots, a_i\}$, where $i \leq n$, we get

$$\{setTuple(t) \mid t \in \mathcal{N}'_B([\![R_0(a_1)]\!]_\varsigma) \times \ldots \times \mathcal{N}'_B([\![R_0(a_i)]\!]_\varsigma)$$
$$\times \mathcal{N}'_B([\![skip]\!]_\varsigma) \times \ldots \times \mathcal{N}'_B([\![skip]\!]_\varsigma)\}$$

where the expression $\mathcal{N}'_B([\![skip]\!]_\varsigma)$ occurs $(n - i)$ times. These expressions represent the else-cases where the guard is not satisfied. According to the definition of $setTuple$ and the semantics of skip rules, this equals

$$\{setTuple(t) \mid t \in \mathcal{N}'_B([\![R_0(a_1)]\!]_\varsigma) \times \ldots \times \mathcal{N}'_B([\![R_0(a_i)]\!]_\varsigma)\}$$

If we assume that $\mathcal{N}'_B([\![R_0(a_i)]\!]_\varsigma) = \mathcal{N}_B(R_0(a_i))$ for any a_i, this equals the semantics of the non-deterministic **do-forall** rule (see Definition (2.17)). Thus, we get

$$\mathcal{N}'_B([\![\texttt{do forall } v \;:\; g(v) \;\; R_0(v) \texttt{ enddo}]\!]_\varsigma)$$
$$= \mathcal{N}_B(\texttt{do forall } v \;:\; g(v) \;\; R_0(v) \texttt{ enddo}).$$

5. **choose** rules:

$\mathcal{N}'_B([\![\text{choose } v : g(v) \ R_0(v) \ \text{endchoose}]\!]_\varsigma)$

$= \ \mathcal{N}'_B(\text{if } [\![g(e)]\!]_\varsigma \text{ then } [\![R_0(e)]\!]_\varsigma)$ 　　　　(according to definition of $[\![.]\!]_\varsigma$)

where e is a nullary external function whose value is chosen non-deterministically from the set $\{a_i \in S_i^A \mid B \models [\![g(a_i)]\!]_\varsigma\}$. Since $B \models [\![g(e)]\!]_\varsigma$, this evaluates to

$\{setTuple(t) \mid t \in \mathcal{N}'_B([\![R_0(e)]\!]_\varsigma) \times \mathcal{N}'_B([\![skip]\!]_\varsigma)\}$

$= \ \{setTuple(t) \mid t \in \mathcal{N}'_B([\![R_0(e)]\!]_\varsigma) \times \{\{\}\}\}$

$= \ \mathcal{N}'_B([\![R_0(e)]\!]_\varsigma)$ 　　　　(according to Definition (5.22))

Since e's value is chosen non-deterministically, this expression is equal to the set

$\bigcup_{a \in \mathcal{RG}} \mathcal{N}'_{B(e \mapsto a)}([\![R_0(e)]\!]_\varsigma)$ where $\mathcal{RG} = \{a_i \in S_i^A \mid B \models [\![g(a_i)]\!]_\varsigma\}$.

If we assume that $\mathcal{N}'_B([\![R_0(e)]\!]_\varsigma) = \mathcal{N}_B(R_0(e))$ for any R_0, it remains to prove that in any expanded state B, for any value a_i

$$B \models [\![g(a_i)]\!]_\varsigma \ \Leftrightarrow \ B \models g(a_i).$$

With the proof of the correctness of term simplification given above, we get $[\![g(a_i)]\!]_\varsigma = Val_B(g(a_i))$ for any boolean term or any first order term $g(a_i)$. It follows that $g(a_i)$ and $[\![g(a_i)]\!]_\varsigma$ are satisfied in the same states and thus

$$\mathcal{N}'_B([\![\text{choose } v : g(v) \ R_0(v) \ \text{endchoose}]\!]_\varsigma)$$
$$= \mathcal{N}_B(\text{choose } v : g(v) \ R_0(v) \ \text{endchoose}).$$

This completes the correctness proof of the transformation algorithm from ASM transition rules into ASM-IL. Due to the simplicity of ASM-IL, the transformation from ASM-IL into the model checker language is very easy.

The next chapters introduce two applications for this general interface: We build on top of ASM-IL an interface to the SMV model checker and an interface to the MDG-package. The former is described in the next chapter. Chapter 7 illustrates this approach using two case studies.

Chapter 6

The SMV Model Checker

Model checking is a powerful means of supporting the analysis of system models in an early stage of development and design. As we discussed in the last chapter, model checking is applicable to ASM models. By means of an interface between an ASM tool framework for developing ASM models and a given model checker, we can provide automatic tool support for analysing ASM.

In the last chapter, we introduced our general interface language ASM-IL for the ASM Workbench (ASM-WB, c.f., [Cas00] and see also Section 2.3) which is a framework for editing, parsing, type-checking, and simulating ASM-SL models. Given the ASM-IL format, this chapter shows how we build a transformation to the input language of the model checker SMV ([McM93]) on top of it. This work completes the interface between the ASM-WB and the model checker SMV. SMV has been chosen as a typical representative of a class of model checkers based on transition systems. It could be easily replaced by another model checker that is based on a similar approach, e.g., SVE [F+96] or VIS [Gro96].

Our transformation tool also supplies SMV with a higher-level modelling language, namely ASM-SL. This facilitates the modelling task by allowing the use of more complex data types, in particular n-ary functions. Functions generalise the classical notion of state variables and allow parameterisation. Therefore, ASM models, in comparison with similar SMV models, are more concise yet more general by means of using parameters. Also, ASM models can be scaled up more easily than SMV models.

The process of transforming an ASM model into the corresponding SMV model is fully automatic. The implementation of the algorithms introduced in this work is available for the ASM-WB, Version 0.99 (University of Paderborn, 1997-99) and SMV 2.5.3.1b (Carnegie Mellon University, November 1998).

In the following section, we give an overview of the SMV tool with its features and describe the SMV input language. This language is the target language of the transformation. Based on this, we compare ASM and the SMV language. In the next section, we introduce how to extend the generic interface (as described in Chapter 5.2) to complete the transformation algorithm from ASM into the SMV language.

We evaluate the use of SMV for checking ASM models by means of two case studies: the FLASH cache coherence protocol as a distributed architecture of communicating processes, and the Production Cell as an embedded system. Both case studies have their own special intricacies and involve a particular solution. These are introduced in the next chapter.

6.1   The SMV Tool

SMV is a tool for symbolic model checking of CTL formulas (see Section 3.3.3). It was introduced by McMillan in [McM93]. Since this time, the original implementation has been reworked and further developed in various directions: McMillan has developed his own version known as Cadence SMV (see [Mil]); at the Carnegie Mellon University the original version of SMV has been maintained and as a re-implementation and extension the model checker NuSMV ([CCGR99]) is launched. The CMU furthermore promotes the framework SyMP, which is a combination of the model checker SMV and some decision procedures for satisfiability (SAT solvers) ([Ber99]). SMV can be seen as the dinosaur of the family of symbolic CTL model checkers. As visible through the literature, the tool is widely used and often compared with other tools.

The SMV tool takes a system specification and a requirement specification as input (see Figure 6.1). Based on this, it checks if the system satisfies the requirements (c.f., the introduction given in Section 3.3). The system specification has to be formalised as a transition system in the SMV language. The requirements have to be formalised in CTL. If the system model violates the requirements, the SMV tool computes a counterexample. A

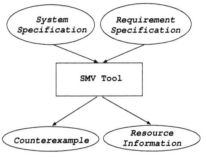

Figure 6.1: The SMV model checker

counterexample is a possible run of the system which is given as a sequence of states. These states are specified by means of the particular evaluation of the state variables. This depiction is shortened such that only those state variable evaluations that are changed in the last step are listed. State variables which are not listed have not changed their value. A counterexample is very useful for debugging the system model under investigation. As a further output, SMV provides information about resources that are used: the size of the internally computed BDD structure and the computational effort of the checking procedure in time and space. This allows the user to compare the effort used for different modelling approaches and also to compare the resources used by the SMV tool with resources needed by other model checkers for checking a similar model.

Simulating the Counterexample

In order to support the debugging process, we aim at a bidirectional interface between the ASM-WB and the SMV tool. That is, the counterexample that is output by the SMV tool should be re-transformed into the ASM-WB. This re-transformation allows for simulating runs of the investigated ASM model.

Although this facility is not yet implemented, it is conceptually easy to realise: The simulator of the ASM-WB computes in each step the next state of a run according to the transition rules. If the model behaves non-deterministically by means of external functions, the simulator asks the user for input. As a response, the user has to determine the behaviour of the external functions in the next step.

Since a counterexample given by the SMV tool specifies the evaluation of each state variable, we can use the counterexample as input for the simulator. By means of a filter mechanism, we choose those state variables that result from transforming the external functions. The evaluation of these (former external) state variables can be used to trigger the simulation run. As a consequence, this run leads to the erroneous state that is indicated by the counterexample. The simulation of the counterexample in the ASM-WB could be executed automatically.

6.1.1   The SMV Language

The SMV tool provides the user with its own specification language, the SMV language. Its syntax and semantics is defined in [McM93]. We recall here the basic features only[1]. Basically, this language facilitates the specification of the system model and its requirements.

The system model.   The system model is a transition system with a set of states, a subset of initial states, and a transition relation that specifies the behaviour. A system is defined as a module (beginning with the keyword MODULE) that may consist of several submodules.

A module describes a synchronous or asynchronous process. It is an encapsulated collection of declarations (given as VAR, INIT, ASSIGN, FAIRNESS or SPEC declarations, see below) that may depend on specific parameters. These parameters have to be instantiated within the calling module (i.e., the parent). In particular, each process has an implicit parameter **running**, a boolean variable that indicates if the process is active or not.

The state (or the state space) of a module is given by declaring state variables of a certain type. This declaration is headed by the keyword VAR. Suitable types that can be used in the declaration part are the set of Boolean values (where *false* is represented as 0 and and *true* as 1), an enumeration of constants, or modules that specify synchronous or asynchronous sub-processes (for the latter, the keyword **process** is used).

The set of initial states can be specified as a conjunction of equations that specify the initial evaluation of state variables. This boolean formula is headed by the keyword INIT. If the initial value of a state variable is not specified, any value (of its declared type) is a suitable initial evaluation. The result is a *set* of initial states rather than a single initial state.

The transition relation of a module is specified in the section headed by the keyword ASSIGN. It contains a list of value assignments to the state variables. The semantics of assignments is given as a next-step evaluation: all assignments are applied simultaneously in one step and thus define the next state by means of a new evaluation of state variables. The state variables on the left-hand side of an assignment are wrapped by the function **next** to specify the corresponding next state variable. The right-hand side of an assignment is an expression. Expressions comprise operators of boolean logic, some operators for integer arithmetic (e.g., addition, subtraction, or multiplication), and a **case** construct which lists various cases for an assignment guarded by a condition. Each case comprise a condition and a value or an expression that should be assigned in case the condition is satisfied. In every run of the model, only the *first* case in the case expression whose condition is satisfied is applied and the state variable is assigned with the corresponding value or expression.

The requirements.   For specifying the requirements to be checked, the SMV language provides the syntax for CTL. It comprises boolean operator for negation (!), conjunction (**&**), disjunction (|), logical implication (**->**), and logical equivalence

---

[1]Our description is not exhaustive but restricted to the features that are interesting in our context.

(`<->`) of simple formulas. The temporal operator **X**, **F**, **G**, and **U** are defined as well as the path quantifiers **E** and **A** (c.f., Definition 3.16 on page 29). The requirements are declared under the heading `SPEC`.

Fairness. As another basic feature, the SMV language provides a means for specifying fairness constraints (see Section 3.5) (keyword `FAIRNESS`). A fairness constraint is a CTL formula. It allows the designer to restrict the investigated runs of a module to those which are fair with respect to the given fairness constraints. That is, these constraints are true infinitely often during each run of the *fair* system module.

6.1.2   ASM versus SMV Language

A comparison of ASM and the SMV language leads to the following points:

- Both notations have in common that a next state in a computation is reached by *simultaneous* assignment of new values.

- Both languages have a notion for modules (or processes). In ASM, these are called *agents*. We can model processes that run simultaneously (all processes are active concurrently) or specify an interleaving behaviour (only one process is active at each time). In the SMV language, interleaving semantics is forced by using the keyword `process` when declaring a module component. In ASM, an interleaving semantics can be modelled by means of external functions: If in a multi-agent ASM the parameter *Self* that indicates the currently active agent is declared as an external function, it evaluates to a particular value at each state. That is, no two processes are active at the same time.

- Both languages do not provide a particular model for communication between concurrent modules.

- ASM allows the designer to model in terms of n-ary functions instead of simple state variables. This supports modelling on a high level of abstraction. The data structure in the SMV language is much simpler and leads to more low-level models.

- In ASM, non-determinism is introduce by means of external functions or the `choose` rule. In SMV, non-deterministic behaviour can be modelled through "under-specification": Any state variable without a specified transition behaviour can behave arbitrarily. In this case, the checking algorithm investigates every possible behaviour.[2]

- Fairness assumptions can be stated in the SMV language but not in ASM.

After this general overview of the SMV tool and its input language, we introduce the interface between the intermediate language ASM-IL and the SMV language in the next section. This second transformation step completes the interface from the ASM Workbench to the SMV model checker.

---

[2]Another means to model non-deterministic behaviour is given through the `TRANS` construct which is not used in our approach.

6.2 From ASM-IL into SMV Language

In this section, we introduce a general schema for an effective coding of ASM rules in the SMV language that preserves the semantics. This result is published in [Win97]. The schema yields the basis for a transformation algorithm that is implemented.

6.2.1 Mapping ASM into SMV

The semantics of the SMV language is defined in [McM93]. We use this definition in order to show that the chosen transformation schema is semantically correct. Since the schema is only applied to the intermediate language ASM-IL (we are discussing the second step of the transformation), it suffices to show correctness for simple guarded update rules (introduced in Section 5.2) of the form

$$\textbf{if } guard_{ij} \textbf{ then } loc_i := val_{ij}$$

where each location loc_i is attached with a list of the guard-update pairs contributing in the model (assuming the model contains m guarded-updates for loc_i then $1 \leq j \leq m$). Note that every val_{ij} is a constant value or an expression that contains only *primary operators* (see Section 5.2.1 on page 75). These primary operators are exactly those that are also supported by the SMV language. Analogously, $guard_{ij}$ is a boolean expression that contains primary operators only. All other operators are unfolded and simplified during the first transformation step.

We argue in two steps: Firstly, we show that the transformation of a simple update is correctly transformed. Secondly, we show the correctness of the transformation of a list of simple guarded updates.

Transformation schema for a simple update   The SMV language deals with variables over finite domains. In ASM, we can identify the notion of *locations* with these state variables (see Section 5.2). A location is a pair $loc = (f, Val_B(\bar{t}))$ consisting of a function symbol f and a list of parameter values $Val_B(\bar{t})$ with respect to the expanded state $B = (A, \zeta)$ (see Definition 2.11 on page 16).

In SMV, we model the updating of a variable by an assignment via the next operator $\textbf{next}(\cdot)$, which specifies the value of the variable in the next state. We choose the following schema for transforming a simple ASM update into SMV syntax:

$$\textbf{f}(\bar{\textbf{t}}) := \textbf{t}_0$$
$$\Downarrow$$
$$\textbf{ASSIGN next}(loc) := val$$

with the location $loc = (f, Val_B(\bar{t}))$ and the value $val = Val_B(t_0)$

In [McM93], the semantics for SMV programs is defined as a function $[\![.]\!]smv$ that maps a program fragment to its denotation. The denotation is a set. This set is either a singleton containing a boolean value or given by a special boolean function that yields code for an OBDD, the internal data structure of the SMV tool. This function is represented by $\rightarrow$, the *if-then-else* operator which defines the *Shannon decomposition* of an OBDD (see Section 4.1, equation (4.1)). The semantics of an expression $a \rightarrow (b, c)$ is given as "if a then b else c".

The semantics of an expression in SMV is denoted by itself [3]: $[\![e]\!]smv = e$. The semantics of an assignment containing a next operator is given as follows:

$$[\![\textbf{ASSIGN next}(loc) := val]\!]smv = (\ loc' \in (\textbf{running} \rightarrow ([\![val]\!]smv, loc)\ ))$$

---

[3]Internally, expressions containing non-boolean operators are mapped into a boolean expression by using the *if-then-else* operator and a given set of axioms. Non-boolean values are encoded similarly (for more detail see [McM93], page 80).

where loc' is the value of the location loc in the next state, and **running** is the internal variable that is attached to each SMV module and indicates if the module is currently active. We can read the semantics of assignments as follows: If the module to which the statement belongs is active then loc' evaluates to an element of the set of possible values of the expression $val = [\![val]\!]_{smv}$, otherwise it remains as it is.

In the semantics of ASM, every transition rule can fire in every state (if it is not guarded), i.e., **running** is always true. If a transition rule fires then the value of the location in the next state is evaluated to $val = Val_B(t_0)$. Therefore, the schema for transforming ASM updates into SMV assignments is semantically correct.

Transformation schema for a list of guarded update rules   The transformation schema for a list of guarded update rules for a particular location loc_i uses the case construct of the SMV language. Each of the simple rules results in one case: the guard is mapped into the leading condition, and the corresponding value (or the expression) is mapped into to corresponding right hand side of the case.

In SMV, a variable behaves non-deterministically if no update is specified. To ensure that a variable will not be changed if none of the guarded update rules fires, we have to state this explicitly since otherwise the variable can take any value from its range. This is done automatically by adding a *default case* which is guarded by the condition *true* (represented by 1 in the SMV language). If none of the guards is satisfied, the location evaluates in the next state to the same value.

$$
\begin{aligned}
&\textbf{if } guard_{i1} \textbf{ then } loc_i := val_{i1}\ , \\
&\qquad\vdots \\
&\textbf{if } guard_{im} \textbf{ then } loc_i := val_{im}
\end{aligned}
\quad \Rightarrow \quad
\left\{
\begin{aligned}
&\textbf{ASSIGN} \\
&\quad \textbf{next } (loc_i) := \\
&\qquad \textbf{case} \\
&\qquad\quad guard_{i1} : val_{i1}\ ; \\
&\qquad\qquad\vdots \\
&\qquad\quad guard_{im} : val_{im}\ ; \\
&\qquad\qquad\quad 1 : loc_i \quad ; \\
&\qquad \textbf{esac}\ ;
\end{aligned}
\right.
$$

Similar to the simple assignment, the semantics of an assignment containing a case block is given as a nested application of *if-then-else* operators. The semantics is shown below:

$$
\left[\!\!\left[
\begin{aligned}
&\textbf{ASSIGN next } (loc_i) := \\
&\quad \textbf{case} \\
&\qquad guard_{i1} : val_{i1}\ ; \\
&\qquad\quad\vdots \\
&\qquad guard_{im} : val_{im}\ ; \\
&\qquad\qquad 1 : loc_i \quad ; \\
&\quad \textbf{esac}\ ;
\end{aligned}
\right]\!\!\right]_{smv}
=
$$

$$
\begin{aligned}
=\ &[\![guard_{i1}]\!]_{smv} \rightarrow (\ (loc_i' \in (\textbf{running} \rightarrow ([\![val_{i1}]\!]_{smv}, loc_i))), \\
&\quad [\![guard_{i2}]\!]_{smv} \rightarrow (\ (loc_i' \in (\textbf{running} \rightarrow ([\![val_{i2}]\!]_{smv}, loc_i))), \\
&\qquad\vdots \\
&\quad\quad [\![guard_{im}]\!]_{smv} \rightarrow (\ (loc_i' \in (\textbf{running} \rightarrow ([\![val_{im}]\!]_{smv}, loc_i))), \\
&\qquad\qquad\qquad\qquad\qquad (loc_i' = loc_i)...)
\end{aligned}
$$

If $guard_{ij}$ is the first condition in the case block that evaluates to true then the value of loc_i' is a member of the set of possible values of $[\![val_{ij}]\!]smv$ whenever the module is active, i.e., **running** is true. Given modules that are permanently active, loc_i' is evaluated to $[\![val_{ij}]\!]smv = val_{ij}$ whenever $[\![guard_{ij}]\!]smv = guard_{ij}$ is true. If none of the guards is satisfied the default case is evaluated. Since the condition of the default case is always true, the location keeps its old values in this case.

With the same argument as in the case above, we can state that this correctly reflects the semantics of ASM in the general case. However, we have to exclude the case in which the ASM model contains inconsistent updates, i.e., updates that address the same location and can be fired in the same state.

With our transformation, inconsistent updates lead to a case block that contains cases with leading conditions that do not exclude each other. That is, two guards $guard_{ij}$ and $guard_{ik}$ ($j \neq k$) may be satisfied in the same state. According to the semantics of the SMV language, the tool will always choose the first case for which the condition is satisfied and apply the corresponding assignment. This does not coincide with the ASM semantics. Therefore, inconsistent update should be excluded.

The proposed transformation schema provides a set of assignments of the form shown above. Each of them assigns a new value to one location only. We can see that this set has exactly the same semantics as the initially given ASM rule if we assume that the rule does not contain inconsistent updates.

Modules

In our transformation, we assemble the set of assignments in a module `behaviour`. The declaration part for the state variables (i.e., the contributing locations) is gathered in a separate module *state*. This module contains also the `INIT` declaration, which is a conjunction of all initial evaluations that are specified in the ASM model. The overall module *main*, which describes the whole system, declares both modules where the module `state` is used as a parameter for the module *behaviour*. It also contains the specification of the requirements in CTL logic (see discussion below). Module *main* has the following appearance:

```
MODULE state
VAR
    ... (declaration of locations)
INIT
    ... (initialisation of locations)


MODULE behaviour (s)
ASSIGN
    ... (assignments of locations)


MODULE main
VAR
    s  : state;
    tr : behaviour (s)
SPEC
    ... (requirements as CTL formula)
```

The same module schema is also used by Havelund and Shankar in [HS95]. We choose this schema for the following reason: If we want to treat ASM that consist of several modules or agents, this schema allows us to separate globally and locally accessible state variables. Only the global variables are declared in the module **state**, the local variables are declared in the sub-modules. This enables us to simulate communication via *global variables* that can be accessed by each module.

We do not map the module concept of ASM into modules of SMV. A multi-agent ASM that uses the parameter *Self* for parameterising the transition rules is mapped into a flat structure of simultaneously applied assignments. We use an interleaving semantics for the concurrent behaviour of the agents. For this approach, the parameter *Self* has to be modelled as an external function in the ASM model. This external function is transformed into an ordinary state variable in the SMV code that changes its value non-deterministically. The parameterised ASM rules are unfolded (according to the transformation algorithm introduced in Section 5.2) in such a way that the evaluation of *Self* turns into a guard for applying a particular assignment: if *Self* equals agent a_i then apply assignments for a_i. Since *Self* has only one value at a time, the resulting SMV model represents an interleaving semantics.

As a consequence, the transformation algorithm is very simple and straightforward. We choose this simple mapping due to the fact that the SMV code is not visible to the user anyway. It becomes an intermediate format. That is, modularisation on the level of SMV code is not helpful for validating the model.

Still, to maintain the modularisation could be useful if we want to decompose the problem at hand into smaller independent components. This way, the model checking task could be split into smaller sub-tasks by checking each component separately. In the investigated case studies, however, decomposition was not helpful: In both cases, there are too many interdependencies between the modules (or agents). Almost every state variable that belongs to one module is used in some other modules as well. Therefore, the effort that would be necessary to resolve these interdependencies would lead to components being as large as the entire model.

Another reason for keeping modularity might be given through the size of the resulting OBDD representation of the model. In the case of the FLASH protocol (see Section 7.1), we investigated the behaviour of the SMV tool for both cases, with and without modules. Prototypically, we transformed the multi-agent ASM model into an SMV model of several processes with an interleaving semantics (i.e., we declared the modules as **process**). Comparing the statistics of used resources during the checking process, we find that mapping agents into separate processes leads to a larger OBDD representation. Thus, the implemented transformation, which expects the parameter *Self* to be specified as an external function that is unfolded, is more effective and the resulting SMV model with a flat structure yields a better coding of this example.

Requirements and Fairness

Requirements, i.e., the properties of the system that we want to check, as well as fairness assumptions, have to be specified as CTL formulas. ASM-SL does not yet comprise a proper syntax for specifying in CTL but as future work we are planning to extend the language with the necessary operators.

Currently, requirements and fairness assumptions have to be added manually into the code by using the SMV language.

Apart from the requirement specification and fairness assumptions, the transformation from ASM models into SMV models operates fully automatically. An implementation of the algorithms is available. In the next chapter, we describe our experiences when applying the transformation and running the SMV tool for analysing ASM models. We show in Section 7.1 through the first case study, the FLASH cache coherence protocol, how model checking can nicely support the debugging process. The second case study, the Production Cell, yields some insights into the limitation of this approach. This is discussed in Section 7.2.

Chapter 7

Case Studies

7.1 The FLASH Cache Coherence Protocol

As one case study for model checking ASM by using SMV, we chose the FLASH
protocol [KOH+94]. Our ASM model of the protocol is based on the work of Durand
[Dur98]. For this model, some refinements are necessary in order to fit the model
for the model checking process. After a short introduction to the FLASH protocol,
we describe the ASM model that is derived from the model given in [Dur98] and
motivate our refinements. In the last subsection, we sketch the debugging process
that is supported by transforming the model and model checking with the SMV
tool.

7.1.1 The Protocol

The Stanford FLASH multiprocessor (cf. [KOH+94]) is an architecture for a large
number of distributed processor nodes that share memory. The memory is distributed over the contributing processors nodes, each of which holds some part of
the memory. In the architecture, the processor nodes are interconnected so that the
distributed memory parts are accessible to every node in order that they may read
or write data.

The protocol that organises the communication between the nodes includes support for *cache coherence* of the shared memory. Since the access to data is realised
by means of holding a copy of the data, this copy becomes *invalid* as soon as there
is writing access to the original data the copy is made from. Cache coherence is
satisfied if the protocol guarantees that none of the nodes hold a copy of data that
has changed in the meanwhile.

The distributed memory is partitioned into *lines*, i.e., small pieces of memory
content. Each line is associated with a *home-node* which hosts the part of the
physical memory where the line resides. Whenever a processor node wants to access
a particular line of which it has no valid copy, a *read-* or *write-miss* occurs. As a
consequence the node sends a *request* to the line's home-node.

Every request should be fulfilled as soon as possible by means of sending a valid
copy of the requested line. A node has *shared access* if it holds a copy only for
reading purposes. A node with shared access to a line is called a *sharer* of the line.
Whenever a node wants to write (i.e., change) data, it needs *exclusive access*. A
node with exclusive access to a line is called the *owner* of the line.

Any request that is caused by a read-miss can be satisfied if the line is not
exclusively accessed by another node. If a write-miss occurs the request can be
serviced only after some preparations:

- Any other node that has shared access to the same line must be informed about the requested exclusive access.

- Subsequently, every sharer must invalidate its copy.

- Every sharer of the line that is informed and has invalidated its copy has to send an acknowledgement to the home-node of the line.

Only if an acknowledgement from every sharer has arrived the home-node can grant exclusive access to the line and sends a copy to the requesting node. This node becomes the owner of the line.

Being interconnected, the nodes are able to communicate with each other as described above: They can send messages to request a remote line or send a copy of the needed data to service a request. To provide coherence of the data, additional book-keeping is necessary to prevent simultaneous reading and writing on the same line. Message passing and book-keeping of shared and exclusive access is the responsibility of the protocol that we consider in our ASM model.

7.1.2 The ASM Model of the FLASH Protocol

The ASM model of the protocol in [Dur98] is based on agents. Each agent models one processor node that holds certain memory lines. A set of parameterised transition rules describes the behaviour of a single agent. Its behaviour is determined by the incoming messages, which can be requests, acknowledgements or grants. That is, each rule describes the behaviour for one possible kind of message that can be received. This modelling idea yields a clear structure. (A complete model in ASM-SL syntax that includes also our refinements is given in the appendix, see Section A.2.)

In this subsection, we describe the ASM model of the FLASH protocol as given in [Dur98] with only one minor change that is motivated and explained (see below). This description should help to understand the necessary refinements of the model (given in Section 7.1.3) and the counterexamples that were found by the SMV tool (given in Section 7.1.4). We start with a summary of the domains and functions that describe the state of an agent. We explain the structure of messages and based on this the behaviour of a communicating agent as specified through the ASM transition rules.

Domains and state functions. The domains of the model are given by the following sets:

- **Agent** models the contributing processor-nodes.

- **Line** models the lines of the memory.

- **Type** $= \{get, put, fwdget, swb, nack, nackc, getx, putx, fwdgetx, inv, invAck, fwdAck, rpl, wb\}$ models the type of messages.

- **ReqType** $= \{cc.get, cc.getx, cc.rpl, cc.wb\}$ models the type of newly generated request messages.

- **Phase** $= \{ready, wait, invalidPhase\}$ models the current phase of a request.

- **Status** $= \{excl, shared, invalid\}$ models the status of the access to a line of memory from the view of each agent, i.e., the access status of the copy of the data content the agent holds.

The agent's behaviour depends on several state functions:

- *CurPhase* : `Agent` × `Line` → `Phase`
  models for each agent the phase of a request to a particular line. If a request is started the current phase of the line is *wait*. If the request is serviced, the phase is reset to *ready*. The current phase is *invalidPhase* if the copy has to be invalidated because of an exclusive access.

- *State* : `Agent` × `Line` → `Status`
  models for each agent the access status of a line copy in use.

- *Pending* : `Line` → `Boolean`
  is a flag indicating a current request for a line. Whenever a request is sent this flag is *true*. It becomes *false* when the request is serviced in some way.

- *Owner* : Line → Agent ∪ {*undef*}
  models the agent that currently has exclusive access to a line. If none of the agents has exclusive access to a line the function equals *undef*.

- *Sharer* : `Line` × `Agent` → `Boolean`
  models if a particular Agent has shared access to a particular line, i.e., holds a copy of the line. Thus, *Sharer* models the characteristic function of the set of *sharers* of a line.

Message Structure. In [Dur98], a message is modelled as a six-tuple consisting of the type of the message, the addressed agent, the sender agent, the agent initiating the request (this must not be the same as the sender), the requested line, and the copy of the requested data.

For applying model checking, we have to abstract this model since the requested data has an infinite domain. In our abstraction, a variable *data* of the original ASM model which models the copy of the data in a message is eliminated. This variable does not influence the control flow of the protocol behaviour since it does not control any of the guards in the transition rules. Also, the variable does not appear in the properties to be checked. Thus, *data* is irrelevant and by discarding it we do not lose expressiveness of the model and our checking results. Eliminating irrelevant variables is an abstraction technique that is also suggested in [BH99] (see also the discussion of related work in Section 9.1).

Each component of a message can be addressed by a special selector function. The most relevant information is given with the type of a message, *MessType*, which is an element of `Type` or `ReqType`.

For **shared** access, we distinguish the following message types:

> `get` : requesting a line from its *home*
> `put` : granting a line to the requester (*source* of the request)
> `fwdget` : forwarding a request to an exclusive owner of the line
> `swb` : requesting a write-back of an owned line that is to be shared
> `nack, nackc` : negatively acknowledging the request or forwarded
>   request (`nackc`), if it cannot be performed now.

In analogy, message types related to **exclusive** access are:

> `getx` : requesting a line for exclusive access from its *home*
> `putx` : granting a line for exclusive access to the requester
> `fwdgetx` : forwarding a request for exclusive access the owner of the line
> `inv` : requesting a current *sharer* of the line to invalidate its local copy
> `invAck` : acknowledging the invalidation of the line
> `fwdAck` : owner's granting according to a forwarded shared request.

Incoming messages are requests from remote nodes or from the receiving node itself. In order to simplify the model, we do not distinguish between a requesting node that is also the home-node of the line and other nodes. The special case of requesting a line that is located on the same node is called *intra-node communication*.

Read- and write-miss. A read- or write-miss that causes a request to be sent is arbitrarily generated by means of an external function *produce* : `Agent` $\rightarrow$ `ReqType`$\times$ `Line`. For each agent, this function maps to a pair (t, l) which causes a message of type t to be sent to the home-node of the line l.

The agents' behaviour. The transition rules given in Figures 7.1 and 7.2 specify the behaviour of an agent. Note that in the figures the parameter l models a line. Moreover, sending of a message is specified as a macro. In Figures 7.1 and 7.2, it is denoted as *SendMsg*. This macro is defined as an **import**-rule that imports a new element, which is a message, into the set of messages in transit, *MessInTransit*. The parameters of *SendMsg* specify the components of the new message.

An arbitrarily generated read- or write miss generates a message of a special type with the prefix "**cc.**". Whenever an agent receives such a message, it sends the corresponding request and enters a waiting mode. Releasing an access is handled in the same way: a message of type **cc.rpl** or **cc.wb** indicates that a memory line is not needed any more. If the agent sends a request for releasing a line, the state of the line is invalidated, i.e., the node has no reliable copy of the line any more. The transition rules for sending a request when a read- or write-miss occurs are shown in Figure 7.1 in the upper part.

The transition rules that are related to requests for shared access are given in Figure 7.1 in the lower part. A *get-request* initiates a sequence of message transfers for granting shared access. The responding behaviour is modelled by the transition rule that is guarded by (*MessType* = **get**) in the following way:

(a) The request is negatively acknowledged if another request on the line is already processing, i.e., *Pending(l)* is true.

(b) The request is forwarded to the current owner if there is already an exclusive access, i.e., if *Owner(l)* is not undefined.

(c) The request is granted to the requester if no owner is noted; this is modelled in the else-case.

If there is an owner (case (b) above), the grant for shared access is given by the agent who "believes" itself to be an owner. Moreover, the owner has to release its exclusive copy.

If an agent gets a *forward get-request*, i.e., (*MessType* = **fwdget**) is true, and does not agree to be the owner of the line (*State(l)* is not exclusive) then the request has to be negatively acknowledged. Otherwise, the owner changes its exclusive access into a shared access, sends a put-message to the requester, that gives shared access to the line under request, and sends a message to the home-node which has to write back the exclusive access and add it to the list of sharers, i.e., a *swb-request* is sent.

The negative acknowledgement, indicating that an access cannot be granted, has to be sent to the requester and to the home-node. The requester gets a message of type **nack**. It resets its current phase to *ready* again which enables it for sending new requests. The home-node gets a message of type **nackc**. The request for a

if $MessType = \boxed{\textbf{cc.get}} \wedge CurPhase(Self, l) = ready$
  then $SendMsg(\textbf{get}, home(l), Self, Self, l)$
    $CurPhase(Self, l) := wait$

if $MessType = \boxed{\textbf{cc.getx}} \wedge CurPhase(Self, l) = ready$
  then $SendMsg(\textbf{getx}, home(l), Self, Self, l)$
    $CurPhase(Self, l) := wait$

if $MessType = \boxed{\textbf{cc.rpl}} \wedge CurPhase(Self, l) = ready \wedge State(Self, l) = shared$
  then $SendMsg(\textbf{rpl}, home(l), Self, Self, l)$
    $State(Self, l) := invalid$

if $MessType = \boxed{\textbf{cc.wb}} \wedge CurPhase(Self, l) = ready \wedge State(Self, l) = excl$
  then $SendMsg(\textbf{wb}, home(l), Self, Self, l)$
    $State(Self, l) := invalid$

---

if $MessType = \boxed{\textbf{get}}$
  then **if** $Pending(l)$ **then** $SendMsg(\textbf{nack}, source, Self, source, l)$
                else **if** $Owner(l) \neq undef$
                        then $SendMsg(\textbf{fwdget}, Owner(l), Self, source, l)$
                          $Pending(l) := true$
                        else $SendMsg(\textbf{put}, source, Self, source, l)$
                          $Sharer(l, \textbf{source}) := true$

if $MessType = \boxed{\textbf{fwdget}}$
  then **if** $State(Self, l) = excl$ **then** $SendMsg(\textbf{put}, source, Self, source, l)$
                        $SendMsg(\textbf{swb}, home(l), Self, source, l)$
                        $State(Self, l) := shared$
                else $SendMsg(\textbf{nack}, source, Self, source, l)$
                     $SendMsg(\textbf{nackc}, home(l), Self, source, l)$

if $MessType = \boxed{\textbf{put}}$ **if** $MessType = \boxed{\textbf{swb}}$
  then $CurPhase(Self, l) := ready$ **then** $Sharer(l, source) := true$
      if $CurPhase(Self, l) \neq invalid$ $Sharer(Owner(l), l) := true$
        then $State(Self, l) := shared$ $Owner(l) := undef$
                                      $Pending(l) := false$

if $MessType = \boxed{\textbf{nack}}$ **if** $MessType = \boxed{\textbf{nackc}}$
  then $CurPhase(Self, l) := ready$ **then** $Pending(l) := false$

Figure 7.1: Responding to a request for shared access

if $MessType = \boxed{\textbf{getx}}$
　　then if $Pending(l)$
　　　　　then $SendMsg(\textbf{nack}, source, Self, source, l)$
　　　　　else if $Owner(l) \neq undef$
　　　　　　　then $SendMsg(\textbf{fwdgetx}, Owner(l), Self, source, l)$
　　　　　　　　　$Pending(l) := true$
　　　　　　　else if $(\exists u : Agents)Sharer(l, u)$
　　　　　　　　　then doforall $u : Sharer(l, u)$
　　　　　　　　　　　　　$SendMsg(\textbf{inv}, u, Self, source, l)$
　　　　　　　　　　　$Pending(l) := true$
　　　　　　　　　else $SendMsg(\textbf{putx}, source, Self, source, l)$
　　　　　　　　　　　$Owner(l) := source$

if $MessType = \boxed{\textbf{fwdgetx}}$
　　then if $State(Self, l) = excl$
　　　　　then $SendMsg(\textbf{putx}, source, Self, source, l)$
　　　　　　　$SendMsg(\textbf{fwdAck}, home(l), Self, source, l)$
　　　　　　　$State(Self, l) := invalid$
　　　　　else $SendMsg(\textbf{nack}, source, Self, source, l)$
　　　　　　　$SendMsg(\textbf{nackc}, home(l), Self, source, l)$

if $MessType = \boxed{\textbf{putx}}$　　　　　**if** $MessType = \boxed{\textbf{fwdAck}}$
　　then $State(Self, l) := excl$　　　　　**then if** $Owner(l) \neq undef$
　　　　　$CurPhase(Self, l) := ready$　　　　　**then** $Owner(l) := source$
　　　　　　　　　　　　　　　　　　　　　　　$Pending(l) := true$

if $MessType = \boxed{\textbf{inv}}$
　　then $SendMsg(\textbf{invAck}, home(l), Self, source, l)$
　　　　　if $State(Self, l) = shared$
　　　　　then $State(Self, l) := invalid$
　　　　　else if $CurPhase(Self, l) = wait$
　　　　　　　then $CurPhase(Self, l) := invalidPhase$

if $MessType = \boxed{\textbf{invAck}}$
　　then $Sharer(l, MessSender) := false$
　　　　　if $((\forall a : Agents)(a \neq MessSender \wedge Sharer(l, a) = false))$
　　　　　then $SendMsg(\textbf{putx}, source, Self, source, l)$
　　　　　　　$Pending(l) := false$

---

if $MessType = \boxed{\textbf{rpl}}$　　　　　**if** $MessType = \boxed{\textbf{wb}}$
　　then if $((\exists u : Agents)Sharer(l, u))$　　　**then if** $Owner(l) \neq undef$
　　　　　$\wedge \neg Pending(l)$　　　　　　　　　**then** $Owner(l) := undef$
　　　　　then $Sharer(l, u) := undef$

Figure 7.2: Responding to a request for exclusive access

particular line is not pending anymore, the function *Pending* must be reset to *false*. As a consequence, new requests to this line can be treated.

The behaviour for treating a request of exclusive access is modelled by the transition rules that are shown in Figure 7.2 in the upper part. A *getx-request* initiates a sequence of message transfers for exclusive access. Analogously to a *get-request* described above, an exclusive request (when *MessType* = **getx**) is treated in the following way:

(a) The request is negatively acknowledged if another request is in process already (by sending a **nack**-message).

(b) If the line has already an owner, the request is forwarded to this owner (by sending an **fwdgetx**-message). If this notified node is truly the owner (i.e., its state of the line is exclusive), it has to invalidate its copy and give exclusive access to the requesting node (by sending a **putx**-message). Otherwise, the request is negatively acknowledged.

(c) If the line has no owners but a number of sharers that hold a copy, these sharers have to be informed about the request for exclusive access (by sending an **inv**-message). When getting an **inv**-message, a sharer has to invalidate its copy of the line and must respond with an acknowledgement of invalidation (sending **invAck**-message). When each of the sharers has sent its invalidation acknowledgement the home-node can grant the exclusive access. It sends a **putx**-message to the requester. That is, the **putx**-message is delayed until all **invAck**-messages are received (c.f., transition rule for (*MessType* = **invAck**)).

(d) If no sharer is listed for the line under request, the exclusive access can be granted immediately (by sending a **putx**-message to the requester).

For releasing a shared or exclusive copy from its cache, an agent sends a write_back (**wb**) or a replace message (**rpl**) to *home*. The agent has to be deleted from the list of sharers or must be declared as not being owner anymore (whichever is appropriate). The corresponding transition rules are given in Figure 7.2 in the lower part.

In the next subsection, we describe our refinements of the model that were necessary for applying model checking.

7.1.3 The Refined Model of the Protocol

In the model given in [Dur98], the macro *SendMsg* adds a message to a (possibly infinite) set of messages in transit. The strategy for receiving a message from this set is not specified. For the proof, it is assumed in [Dur98] that the messages are received in the right order.

In order to formalise this assumption and to keep the model finite, we have to refine the model. By this refinement, we restrict the model to exclude those runs in which a message is sent but never received or sent but overwritten by a message that is sent later. That is, we want to exclude messages being received in a different order than they are sent.

Without our refinements, the model is either infinite (in [Dur98] the set of messages in transit is not restricted at all) or yields a wrong counterexample if we restrict the set to be finite. We cannot guarantee liveness if messages may be received in a wrong order.

In order to avoid loss of messages and to guarantee that the order of sending and receiving messages is proper, we add a strategy of synchronisation:

- We introduce a queue for messages in transit for each agent.

- A message can be sent to an agent only if its queue for messages in transit is not full.

- We interleave the protocol behaviour with synchronisation steps.

Each step of protocol communication (specified by the rules given in Figures 7.1 and 7.2) is followed by one step of message transfer. Our refinement introduces an inconsistency in the intra-node communication, i.e., if the home-node sends a request to itself. We solve this problem by extending the specification. The refinement and necessary changes are described in the following subsections. The full specification of the refined model is given in the appendix (see Section A.2).

Queue for Messages in Transit

Instead of handling messages as tuples, we decide to model the components of a message via functions. This avoids unnecessary complexity caused by the unfolding of tuples in the transformation algorithm (see also the discussion in Section 5.2). The infinite universe MessInTransit of [Dur98] is replaced by finite queues, one for each message component. Consequently, a message has only four components because each queue is attached to the agent that is addressed. This agent becomes a parameter for the message components in our model and QLength is an index for the queue (see below).

```
MessInTr: QLength * Agent -> Type,
SenderInTr: QLength * Agent -> Agent,
SourceInTr: QLength * Agent -> Agent, and
LineInTr: QLength * Agent -> Line.
```

The universes QLength, Agent, and Line are sets that are finitely restricted by a maximal index: maxQ, maxAgent, or maxLine. These constants can easily be adjusted in order to scale up the model under investigation.

Sending a message (*SendMsg*) is now specified as appending the message components to the corresponding functions: If n_i is the smallest index for which the queue for messages in transit is empty, we update all message-component functions at this index. We indicate emptiness of a particular cell in a queue by means of a new message type noMess for the message type component. That is, if (MessInTr(n_i, a_i)=noMess) is satisfied and for all indices $n_j < n_i$ it is not, then n_i marks the index for writing the message to be appended. This is specified by means of the following macro:

```
transition AppendToTransit(agent_,(sender_,mess_,source_,line_)) ==
 if MessInTr(1,agent_)=noMess
 then SenderInTr(1,agent_) := sender_
      MessInTr(1,agent_)   := mess_
      SourceInTr(1,agent_):= source_
      LineInTr(1,agent_)   := line_
 else do forall i in { 2..maxQ }
      if MessInTr((i-1),agent_)!=noMess and MessInTr(i,agent_)=noMess
      then SenderInTr(i,agent_) := sender_
           MessInTr(i,agent_)   := mess_
           SourceInTr(i,agent_):= source_
           LineInTr(i,agent_)   := line_
      endif
      enddo
 endif
```

For requests (i.e. message type is `cc.get`, `cc.getx`, `cc.wb`, or `cc.rpl`), we introduce an extra queue for request in transit `MessInTrR`. Analogously, we define dynamic functions that hold one of the components of a request. Sending of a request message is specified by means of the macro `AppendRequestToTransit` that looks similar to the macro shown above.

Transfer of Messages in Transit

Sending of a message by means of appending it to the queue of messages in transit, is only one half of the message exchange. As a second step, we have to specify how the messages in transit are read by the addressed agents, i.e., how messages are transfered (c.f., Figure 7.3).

Each agent processes a *current message* (which is empty if its type equals `noMess`). Again, we model the current message by means of its components: `InMess`, `InSender`, `InSource`, and `InLine` are dynamic functions over domain `Agent`. The current message is delivered by transferring the first element of the transit queues `MessInTr` and `MessInTrR` (see Figure 7.3). Note that requests have lower priority than messages, i.e. a request is passed through only if there is no message in transit left.

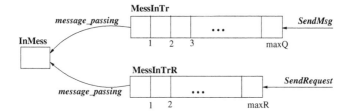

Figure 7.3: Transfer of messages in transit

Processing the current message and delivering a new message involve the same dynamic functions of the model. In order to avoid inconsistencies between the transition rules, we have to interleave both steps: message processing of the agents and transferring of a message. We extend the overall behaviour by means of a sub-step for synchronisation. In the synchronisation step, the messages are transferred to the addressed agent in the proper order.

In an ASM model, introducing a sub-step (to be processed after each state transition) is done in a compositional way: in addition to the ASM for message processing we specify an ASM for the message transfer. That is, we do not have to spread the specification for transferring messages throughout the ordinary transition rules of the agents' behaviour (c.f., Figures 7.1 and 7.2). An overall ASM `main` invokes both "sub-ASM" `agent_behaviour` and `message_passing` in turn:

```
transition main ==
  if toggle = behave
  then agent_behaviour
       toggle := sync
  else (* if toggle = sync *)
       message_passing
       toggle := behave
  endif
```

Using this, we maintain the benefits of the clear and understandable structure of the abstract model that is given by [Dur98].

Inconsistencies for Intra-Node-Communication

Inconsistencies between the transition rules of two agents (i.e., if two agents attempting to update the same location simultaneously) are excluded since we apply an interleaving semantics to the agents' behaviour, and introduce sub-steps for message transfer (as described above). However, inconsistencies may be caused by *intra-node communication*: To keep the ASM model small, we do not distinguish whether the requesting node is also home of the line under request or not. That is, if the home-node of a line wants to have shared or exclusive access to its own line, it has to proceed in the same way as all other nodes, i.e., by sending a get/getx-message to itself.

If the requester is also the home-node of the line, it may happen that two messages are to be sent simultaneously to the same address. Inconsistent updates occur and one of the messages (the first of two updates) will be lost. (In Figures 7.1 and 7.2 the reader may find such situations in the fwdget-rule and the fwdgetx-rule.)

In order to avoid inconsistency, we introduce three new message types that combine two messages: put_swb, putx_fwdAck, and nack_nackc. Whenever the source of the message (i.e. the requester) is equal to the home-node of the line under request, we send one of the combined messages instead of sending two single messages. According to the new message types, we introduce new transition rules that combine the behaviour specified in both of the rules that have to be fired when receiving such messages.

7.1.4   Model Checking the FLASH Model

We use model checking of the transformed ASM model as an evolutionary process of debugging: We edit the ASM model, transform it automatically into an SMV model, run the SMV to check the properties under investigation, simulate the resulting counterexample (if any) with the ASM Workbench, and edit the ASM model again.

Since the debugging process is more efficient if the model checking terminates in a reasonable span of time, we keep our model, in the beginning, as small as possible. We find that even if the model is restricted to two communicating agents and one line of memory, we detect errors in the abstract model as well as in our refinement. In the following, we describe *some* of the detected errors as examples. The errors arise when checking the model for **safety** and **liveness**:

- No two agents have exclusive access to the same line simultaneously.

- Each request will eventually be acknowledged.

- Whenever an agent gets shared access to a line, the line's home will note it as a sharer.

We formalise these requirements in CTL as follows[1]:

---

[1] Although the third specification is rather weak, it yields helpful counterexamples. The disjunction is necessary due to the fact that the sequence, in which the corresponding updates (noting a state to be shared and adding a sharer) occur, may change for different scenarios.

$\bigwedge_{i \neq j}$[AG (!(State(a(i),l)=exclusive & State(a(j),l)=exclusive))]
$\bigwedge_i$[AG (CurPhase(a(i),l) = wait -> AF (CurPhase(a(i),l) = ready))]
$\bigwedge_i$[AG (State(a(i),l)=shared -> AX (Sharer(l,a(i)) = true)
      | Sharer(l,a(i)) = true -> AX (State(a(i),l)=shared))]

Our first counterexample shows simultaneous exclusive access. We simulate the
ASM model and run into an error that can also be found in the abstract ASM
model of [Dur98]:

> Whenever a **putx**-message is sent to grant exclusive access, the addressed
> requester has to be noted as owner of the line. This is specified for the
> **getx**-rule but it is missing for the **invAck**-rule that might also cause a
> **putx**-message to be sent (see also Figure 7.2). The protocol is unsafe
> since simultaneous exclusive access may occur, and written data may be
> lost. (The corresponding counterexample is given in the appendix, see
> Section A.2.2).

Other counterexamples involve the problem of inconsistency (i.e., conflicts caused
by attempted simultaneous updates) on the finite message queue. Although our
data space is restricted to very short queues, we can derive more general remarks
for message passing via queues that concern the finite limitation of a queue itself
(which, in fact, is characteristic for each possible realisation of a queue):

> Each sharer of a requested line has to process the rule for invalidation
> (**inv**-rule). It sends an **invAck**-message to the home-node to acknowl-
> edge the invalidation. When receiving an **invAck**-message, the home-
> node deletes the sender from the list of sharers. If the home-node is a
> sharer too[2], a deadlock may occur if the number of sharers is greater
> than or equal to the length of the message queue: The home-node may
> fail to complete processing the **inv**-rule (see Figure 7.2) when the queue
> is full and sending of a new message to it is not possible (since every
> other sharer may have sent before). The home-node stays busy and can
> not process the incoming **invAck**-rule to clear the queue. In general,
> we found that the size of the message queue must be greater than or
> equal to the number of contributing agents since, in the worst case, each
> agent is a sharer and will simultaneously send an **invAck**-message to
> the home-node. (This scenario is shown by the second counterexample
> given in the appendix, see Section A.2.2).

When queueing messages, we should take into account that sending and processing
a request is not possible in the same instance of time. Independent of the length of
a queue, we should be aware of the following scenario:

> Producing a request for releasing an exclusive access to a line is guarded
> by the condition that the current status shows exclusive access (see
> **cc.wb**-rule in Figure 7.1). When the releasing request is processed, the
> current owner (e.g.,**agent_1**) of the line is reset. Since message passing
> may take time, many things can happen in the mean-time (i.e. after
> appending the request to the queue for requests in transit (**MessInTrR**)
> and before the home-node receives the request). For instance the follow-
> ing scenario can occur: The current owner, **agent_1**, asks for exclusive
> access for a second time. This may be processed and **agent_1** becomes
> the owner again. In the next step, it may happen that another agent
> (**agent_2**) asks for exclusive access and **agent_1** invalidates its copy and

---

[2]This is possible if we allow intra-node communication (see Section 7.1.3).

sets its status to *shared* (see `inv`-rule). The other agent gets exclusive access on the line and becomes the owner (see `invAck`-rule). If the releasing request of `agent_1` is now transfered then the current owner (`agent_2`) is reset although it is not the same owner that asked for release. (The precise scenario is given by the third counterexample in the appendix, see Section A.2.2).

We correct our ASM model in that we repeat the guards of the request-rules (i.e., rules for `cc.get`, `cc.getx`, `cc.wb`, and `cc.rpl`) when the home-node of the requested line receives the requests in transit.

The examples show that helpful borderline cases can be detected more easily by a model checker than by pure simulation. The computational effort for the automated transformation of the ASM model ranges from three to five seconds. The size of the resulting SMV models is given below where the different columns yield the comparing results when scaling up the parameters for the number of agents and lines. The experiments were carried out on an UltraSPARC-II station with 296MHz and 2048 Mb memory, the operating system is Solaris 2.6:

resources used:	2 agents, 1 line	3 agents, 1 line	2 agents, 2 lines
user time/system time:	4.69 s/0.13 s	5687.52 s/0.6 s	17263.2 s/0.86 s
BDD nodes allocated:	70587	1612740	2975127
Bytes allocated:	4849664	37748736	54657024
BDD nodes representing the transition relation:	19261 + 78	288986 + 82	78365 + 96

The output of the SMV tool shows the user time and the system time that is used by the model checking process. The former indicates the time in which the model checking algorithm terminated, the latter shows the time that was spent on system calls. The number of BDD nodes that represent the transition relation is given as a sum because the transition relation is internally represented by the transition and an invariant (e.g., a fairness constraint). The first number in the sum gives the size of the BDD for the transition, the second number specifies the size of the BDD for the invariant. In this example, the second BDD represents the fairness constraint on "fair" evaluation of the variable *Self*.

The variable ordering is determined by the automatic reordering facility that is given by the SMV. Resources used when checking CTL requirements are also listed in the appendix (see Section A.2).

Although checking our model of the FLASH protocol is only feasible for a small number of agents and lines, the results show that the counterexamples yield extremely helpful scenarios for locating errors.

In the next section, we discuss the second case study that we investigated with the SMV tool: the Production Cell. The Production Cell represents an embedded system that, in contrast to the protocol that is discussed in this section, comprises assumptions about the environment behaviour. This allows us to show the limitations of the SMV tool.

7.2 The Production Cell

The Production Cell was introduced in the literature as a case study for comparing different formal notations (see [LL95]). An ASM model of the case study is introduced by Mearelli in [Mea96] (see also [BM97]). Mearelli developed an abstract model and a refined model. Both models describe finite systems since the domains are regarded as sets of discrete, rather than continuous, values. Therefore, the models are suitable for applying model checking. In the first of the following subsections, we describe the case study and the ASM model.

Since the Production Cell is an embedded system, the environment heavily influences the control flow. In order to exclude unexpected or non-reasonable behaviour, additional assumptions on the environment are necessary. These assumptions have to be added to the model we want to treat by the model checker. In the second subsection, we describe in which way we model the environment assumptions and discuss the consequences of our experiences for modelling environment assumptions in general.

The third subsection summarises the model checking procedure and its results with respect to the additional environment model.

7.2.1 The Case Study

The case study was launched by the Forschungszentrum Informatik in Karlsruhe. It describes a model that represents an actual industrial facility of a metal-processing plant, called the Production Cell.

The Production Cell is a system for controlling a plant for processing metal blanks. Figure 7.4 illustrates the Cell with its contributing components: feed belt, deposit belt, press, robot, elevating rotary table, and a crane. The blanks to be processed by the press are conveyed by the feed belt. A robot takes each blank from the feed belt and places it into the press. The arm has to withdraw from the press in order to avoid crashing when the press starts to form the blank. After processing, the press opens again and the robot takes the forged blank from the press and put it on the deposit belt.

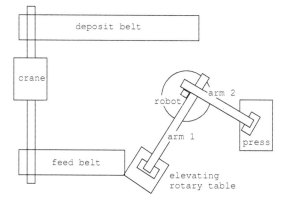

Figure 7.4: The Production Cell

To optimise the process, the robot has two arms, **arm_1** and **arm_2** (see Figure

7.4). **arm_1** rotates counterclockwise to load the press with a new blank. **arm_2** is responsible for unloading the press after forging a blank. It also rotates counterclockwise when unloading. Robot and press are active simultaneously. While the press is forging a blank, **arm_1** picks up another blank that should be processed next. Since **arm_1** is not placed on the same vertical level as the feed belt, the elevating rotary table has to compensate for the difference by means of vertical movement. Similarly, the press has to work on different levels as well. In the model, three states are distinguished for the press: open for loading by **arm_1** (the upper arm), open for unloading by **arm_2** (the lower arm) and closed (i.e., forging the blank). In order to avoid user interaction, the Production Cell is described as a cyclic process. Each blank is taken by the crane from the deposit belt (after being processed) to the feed belt again. A photo-electric cell informs the crane when a blank arrives at the end of the deposit belt. A similar cell is also installed at the end of the feed belt so that the arrival of each blank is detected.

The Production Cell describes an *embedded system*. An embedded system consists of a physical process and a controller which can be a piece of software or hardware. The physical process is called the *environment*. It comprises sensors which measure the behaviour of components of the plant (e.g., the photo-electric cell that recognise a blank approaching on the belt) and actuators that trigger the movement of mechanical parts of the plant (e.g., starting or stopping the motor of the press or the feed belt).

The controller reacts on the given situation in the environment by reading the sensor values. It controls the plant by writing output to the actuators. Figure 7.5 sketches a typical architecture of an embedded system such as the Production Cell. Sensor values are input for the controller and actuator values are its output.

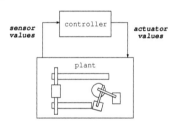

Figure 7.5: Architecture for Embedded Systems

Between the controller software and the physical environment, a compile step is needed that translates the physical signals into the data structures that can be interpreted by the software and vice versa. In most cases, formal specifications have to abstract from these interface issues as well as from a precise description of the environment (i.e., the physical plant). Instead, the model of the environment (including the interface issues) is reduced to a set of (logical) assumptions about its behaviour.

7.2.2   The ASM Model of the Production Cell

The ASM model of the Production Cell is described in detail in [Mea96]. In this section, we present a fragment of the Production Cell only. The full model in ASM-SL syntax is given in the appendix (see Section A.3).

106

The Production Cell is modelled on two levels of abstraction: the ground model and the concrete model. The former models the behaviour of the components by means of states. Whenever it is in a particular state, the system executes the corresponding action. That is, the current state guards the transition rule that is to be fired. In the refined model, the possible states are specified more precisely. Each state is determined by the evaluation of the contributing *sensor variables*. These sensor variables are modelled as external unary functions. The output values for the actuators are given through the *actuator variables*. In the ASM model, these are modelled as unary dynamic functions.

The correctness of the ground model is based on a number of abstract assumption on the behaviour of the environment. We choose the concrete model (which is still small enough to be treated by a model checker, see below) which implements some of the necessary assumptions as the model to work with.

Domains. All domains of the ASM models are finite. We distinguish two different kinds of enumerations:

- Switching of motors: Different domains are introduced for modelling how particular motors in the cell are controlled (called SWITCH_1 to SWITCH_6 in Section A.3); for instance, the motor for the feed belt is simply turned *on* or *off*; the motor of the robot, in contrast, is either *idle* or running *clockwise* or *counterclockwise*.

- Sensor values: Two domains are used to describe the position of the two robot arms in an abstract way: the model distinguishes five positions for arm_1 (SENSOR_1) and only four for arm_2 (SENSOR_2). A third domain for sensor values is used to model the position of the vertical gripper of the crane (SENSOR_3).

Most of the dynamic and external functions, however, model boolean variables, i.e., simple flags (e.g., the sensor value of the photo-electric cell that is mentioned earlier).

Accordingly, the ASM model is finite on both levels of abstraction. In fact, the models can be transformed automatically into the corresponding SMV model even without unfolding of n-ary functions (see Section 5.2) since all functions are unary, i.e., they have the form of variables already.

Modules. The components of the Production Cell are modelled in [Mea96] as *modules* that run in parallel. A module consists of a set of transition rules that describe the component's behaviour. However, there is no distinction between functions that are only accessible within a module (i.e., functions that are *local*), and functions that are accessible by other modules as well (*global* functions). Thus, the module structure turns out to be a flat structure of transition rules without local name spaces, i.e., every function is globally known. The transition rules, according to the ASM semantics, fire simultaneously.

Every component in the Production Cell executes a sequence of actions depending on the situation in the cell. This can be easily modelled by simple conditional update rules: in a particular situation (modelled by the guard of the rule) the component does something (modelled by the corresponding set of updates).

To give an example, we describe the transition rules of the concrete model that describe the behaviour of the feed belt (see Figure 7.6). The rules specify the behaviour of the belt in three different states: *NormalRun*, *Stopped*, and *CriticalRun*. These states are specified through the status of the feed belt motor (being on or off) and the flag *Delivering* indicating if a blank is delivered to the table.

$\boxed{\text{FB\_NORMAL}}$

if $NormalRun \land PieceInFeedBeltLightBarrier$
  then $FeedBeltFree := true$
      if $TableReadyForLoading$
        then $Delivering := true$
        else $FeedBeltMot := off$

$\boxed{\text{FB\_STOPPED}}$

if $Stopped \land TableReadyForLoading$
  then $FeedBeltMot := on$
      $Delivering := true$

$\boxed{\text{FB\_CRITICAL}}$

if $CriticalRun \land \neg PieceInFeedBeltLightBarrier$
  then $Delivering := false$
      $TableLoaded := true$

where $NormalRun \equiv FeedBeltMot = on \land \neg Delivering$
        $CriticalRun \equiv FeedBeltMot = on \land Delivering$
        $Stopped \equiv FeedBeltMot = off$
        $TableReadyForLoading \equiv TableInLoadPosition \land \neg TableLoaded$
        $TableInLoadPosition \equiv BottomPosition \land MinRotation$
                     $\land TableElevationMot = idle$
                     $\land TableRotationMot = idle$

Figure 7.6: Transition Rules for the Feed Belt

The original model heavily uses macro definitions, like *TableReadyForLoading* for instance (see Figure 7.6). For transforming the model into the SMV language, we expand these macro definitions and replace the appearance of the macros in the rules with the equivalent expressions. In the given fragment of the model (in Figure 7.6), it is not visible that some functions are accessed in more than one module (e.g., *TableElevationMot* or *TableRotationMot*). The function *TableLoaded* is even updated in two different modules, the feed belt and the robot.

Although the behaviour of each component appears to be quite simple, the behaviour of the whole system is not so easily validated. Since all transition rules fire simultaneously, the possible interference is difficult to analyse. Model checking appears to be a helpful means for checking if the *concurrent* behaviour of the components is correct. However, in order to apply model checking, we have to specify the behaviour of the environment. This is discussed in the next section. Our investigations with the SMV tool are summarised in Section 7.2.4.

7.2.3 The Environment Assumptions

In the given ASM model of the Production Cell, the environment is represented by a list of external functions (see Figure 7.7, and also the ASM-SL model in the appendix in Section A.3). Each of the external functions represents a sensor variable that influences the behaviour of the Production Cell.

During the transformation from ASM into the SMV language, external functions are mapped into state variables. Since these state variable are not controlled by

```
external function PieceInFeedBeltLightBarrier :BOOL
external function MaxRotation :BOOL
external function MinRotation :BOOL
external function TopPosition :BOOL
external function BottomPosition :BOOL
external function Arm1Ext :SENSOR_1
external function Arm2Ext :SENSOR_1
external function Angle :SENSOR_2
external function TopPositionPress :BOOL
external function MiddlePositionPress :BOOL
external function BottomPositionPress :BOOL
external function ForgingComplete :BOOL
external function PieceInDepositBeltLightBarrier :BOOL
external function GripperOverDepBelt :BOOL
external function GripperOverFeedBelt :BOOL
external function GripperVerticalPos :SENSOR_3
```

Figure 7.7: External functions that model sensor variables

the system, their behaviour is unspecified in the resulting SMV model. The model checker has to check each possible behaviour. This leads to counterexamples that show not intended behaviour of the environment. We call such an output a *wrong counterexample* (see also the discussion in Section 5.1).

In order to avoid wrong counterexamples in the case of embedded systems, we have to model assumptions that are stated on the environment. For the Production Cell, these assumptions concern the length of the belts (they have to be finite) and the mechanical behaviour of the motors: They are not distorted by mechanical faults and perform at a constant speed which is reached instantaneously after being switched on.

These assumptions are fairly minimal, and yield only a small restriction on the behaviour of the sensor values. In particular, they do not provide any hints about the *duration* of an environment reaction. For instance, the duration for transporting a blank on the belts can be arbitrary. We only assume that if a blank is put onto the belt then *eventually* it will reach the end of the belt if the motor is not stopped for ever. To model this assumption properly, we need a temporal logic.

Since the SMV tool does not support temporal logic for modelling system properties, the environment assumptions have to become a part of the system model. Therefore, we have to model the environment in terms of ASM transition rules[3]. For instance, the behaviour of the sensor variable *PieceInFeedBeltLightBarrier* may toggle in one step if the feed belt motor is running, the feed belt is not free, and the piece is not yet delivered (see the ASM transition rule in Figure 7.8).

A full model of an environment specification is given in the appendix (see Section A.3.2). The model checking results must always be interpreted with respect to the added environment model. If the model checker indicates that the checked properties are satisfied, they still might be violated for a different environment. Errors can be overlooked that appear only in cases that do not match the chosen environment model.

---

[3]Note that through adding these rules the external functions turn into dynamic functions since they are now controlled by the ASM model.

> **if** $FeedBeltMotor = on \land \neg(FeedBeltFree) \land \neg(Delivering)$
> **then** $PieceInFeedBeltLightBarrier := true$
> **else if** $FeedBeltMotor = on \land Delivering$
> **then** $PieceInFeedBeltLightBarrier := false$

Figure 7.8: ASM model for the sensor variable `PieceInFeedBeltLightBarrier`

Alternative approaches for the environment model

Our chosen environment model is deterministic since we fix the duration of reactions. As a result, we get a system model that is too restrictive. Alternatively, we could have modelled non-determinism by means of introducing an external function that determines the number of steps for modelling a non-fixed duration of reactions. This approach would provide *unbounded non-determinism* for the environment behaviour if the external function is modelled as an integer value. However, an external function over an infinite domain cannot be treated by the model checker.

Therefore, we have to restrict the number of choices. For example, a piece on the feed belt may reach the end of the belt in one, two or three steps (c.f., Figure 7.8). However, this model does not completely catch the unbounded non-deterministic behaviour either, which was originally assumed. We cannot guarantee that the critical cases are included in the chosen subset. A crash may occur if the feed belt needs four or five steps for transporting a blank.

7.2.4   Model Checking the Production Cell Model

For model checking the ASM model of the Production Cell, we investigate the safety requirements of the system as they are listed in [LL95] and [Mea96]. Also, we argue that the liveness property can be formalised and checked in an abstract way.

The Safety Requirements of the Production Cell

The safety requirements of the Production Cell are given as properties of its components, but the simplicity of the system allows us to check the system as a whole. We specify each property as a single formula and form the conjunction of this set of formulas in order to get the requirement specification. By means of an example, we show how to formalise the requirements in CTL. The full CTL specification of all requirements of the Production Cell is given in the appendix (see Section A.3.3).

Safety of the Press   As an example, we describe the formalisation of the safety requirements for the press. These requirements are informally given in [LL95] through the following two points:

1. *"The press is not moved downward if it is in the bottom position; it is not moved upward if it is in the top position."*
   This is easily described using the state variables that model the motion of the press motor and the oracle functions for the position of the press: if the bottom position is reached, the press motor should not move downward; similarly, if the top position is reached, the press motor should not move upward.

   We have to bear in mind that the condition of reaching the top or bottom position is also a condition for switching the motor to idle. Therefore, we profit by the next step quantification of CTL stipulating that whenever the boundary is reached in *all possible next states*, the motor should not move the press in this direction[4]. We get:

$$\mathbf{AG}\,(bottomPositionPress \Rightarrow \mathbf{AX}\,(PressMot \neq down))$$
$$\wedge\;\mathbf{AG}\,(topPositionPress \Rightarrow \mathbf{AX}\,(PressMot \neq up))$$

2. *"The press only closes when there is no robot arm inside it."*
The press is closing when the motor is running *up*. In this case, neither **arm_1**
nor **arm_2** should ever be in the press. In terms of the ASM press model, the
arms are in the press if the angle has the value *Arm1ToPress* or *Arm2ToPress*
and the arm is completely extended. This leads to the following formula:

$$\mathbf{AG}\,(PressMot = up$$
$$\Rightarrow \neg(Arm1Ext = Arm1IntoPress \;\wedge\; Angle = Arm1ToPress)$$
$$\wedge\;\neg(Arm2Ext = Arm2IntoPress \;\wedge\; Angle = Arm2ToPress))$$

Similar to the example of the press, we formalise the safety requirements for the
other components (see the appendix, Section A.3.3, for a full listing).

The behaviour of the external functions which model the environment has to be
specified as well in order to reflect the *Cell Assumption* (see [Mea96]) that requires
a reasonable system environment.

The model checker concluded that the conjunction of all safety formulas is
satisfied by the SMV model of the Production Cell with respect to the environment
model that is added. The output shows the resources used:

```
resources used:
user time:  3.8 s, system time:  1.03333 s
BDD nodes allocated:  43529
Bytes allocated:  18087936
BDD nodes representing transition relation:  25718 + 1
```

The Liveness Property of the Production Cell

Liveness of the Production Cell is formulated as *"Every blank inserted into the
system via the feed belt will eventually be dropped by the crane on the feed belt again
and will have been forged"*. Mearelli proved in [Mea96] the liveness property with
reasoning about the blanks that are being passed the system. For applying model
checking, however, this approach is not feasible since blanks are not represented in
our model. We cannot express the liveness properties in terms of blanks. Therefore,
we base our approach on a different proof idea.

Liveness, in the most general sense, means that *eventually something good will
happen* (see e.g. [CMP92]). The good thing that should happen in our case is that
every blank will return to the feed belt and will have been forged. Since the overall
process is a cycle which consists of a sequence of steps, we have to check that it is
always true that the next step in the process cycle will eventually be executable. If
this requirement is satisfied no deadlock will occur.

We split the cyclic process into a sequence of actions for processing one blank.
For the liveness property, it suffices to show that a blank is treated by every component of the cell in this sequence of actions: It should be transported by the feed
belt to the elevating rotary table, should be moved by the robot to the press, and
from there to the deposit belt, where the travelling crane puts it back to the feed
belt. Then the cycle of the process is closed.

---

[4]In the real-life system, we have to require that the boundary be set with a sufficient margin
in order to be able to stop the motor at a safe distance to prevent a collision.

Treating a blank is described in an abstract manner: e.g. a blank is transported by the feed belt if the feed belt changes from *NormalRun* to *CriticalRun*. In Figure 7.9, we identify the pairs of actions which are control-critical moments in the process cycle of the Production Cell.

FeedBelt is in *NormalRun* ⤳	*FeedBelt* is in *CriticalRun*
FeedBelt is in *CriticalRun* ⤳	*Table* is stopped in load position
Table is stopped in load position ⤳	*Table* is stopped in unload position
Table is stopped in unload position ⤳	*Robot* has unloaded *Table*
Robot has unloaded *Table* ⤳	*Press* is *OpenForLoading*
Press is *OpenForLoading* ⤳	*Robot* has loaded the press
Robot has loaded the press ⤳	*Press* is *ClosedForForging*
Press is *ClosedForForging* ⤳	*Press* is *OpenForUnloading*
Press is *OpenForUnloading* ⤳	*Robot* has unloaded the press
Robot has unloaded the press ⤳	*Robot* has loaded *DepositBelt*
Robot has loaded *DepositBelt* ⤳	*DepositBelt* is in *NormalRun*
DepositBelt is in *NormalRun* ⤳	*DepositBelt* is in *CriticalRun*
DepositBelt is in *CriticalRun* ⤳	*Crane* has unloaded *DepositBelt*
Crane has unloaded *DepositBelt* ⤳	*Crane* has loaded *FeedBelt*
Crane has loaded *FeedBelt* ⤳	*FeedBelt* is in *NormalRun*

Figure 7.9: Pairs of actions that describe progress

We formalise progress in terms of these pairs of actions by means of CTL formulas of the form

$$\mathbf{AG}\,(action\_executable \Rightarrow \mathbf{AF}\,(next\_action\_executable))$$

Following the semantics of CTL, this is satisfied if it is always (in every path in every state) the case that once *action* is executable then eventually *next_action* will be executable. We consider all pairs of actions and their successor actions as given above and get a set of CTL-formulas of this form. The liveness property is considered to be the conjunction of all these progress formulas. (The full specification of the liveness property is given in the appendix in Section A.3.4.)

Moreover, to ensure that *action_executable* will be reached at all, we have to verify that *eventually the action will be executable*: ($\mathbf{AF}\,action\_executable$) has to be checked for every first action in the set of action pairs.

We checked our specification of liveness against the SMV model of the concrete system of the Production Cell. We found that the system satisfies the property if (and only if) the initial condition guarantees that there are at least two blanks in the system. Otherwise, the robot cannot operate properly. Due to the optimised design of the Production Cell, the robot, whenever it puts a blank into the press with one arm, it also takes a forged blank from the press by means of the other arm. Since we do not model an operator that puts new blanks on the feed belt, we have to change the initial condition in order to get more than one blank within the system (see the Insertion Priority Assumption and the proof of the Strong Performance Property in [BM97]). We specify the initial states[5] as given in Figure 7.10. This initial condition models the situation that seven blanks are in the Production Cell (where two blanks are lying on each belt). If we add more blanks (e.g., by setting *Arm1Mag* to *on*), the model checker finds a deadlock.

---

[5] With respect to the model given in [Mea96], we only keep the initial values for *Angle*, *Arm1Ext*, *Arm2Ext*, *Arm1Mag*, *Arm2Mag*, and *PressLoaded*. For every other function in Figure 7.10, we change the initialisation.

$$
\begin{array}{rcl}
FeedBeltMot & = & idle \\
FeedBeltFree & = & false \\
PieceInFeedBeltLightBarrier & = & true \\
Angle & = & Arm1ToTable \\
Arm1Ext & = & retracted \\
Arm2Ext & = & retracted \\
Arm1Mag & = & off \\
Arm2Mag & = & off \\
TableLoaded & = & true \\
PressLoaded & = & true \\
DepBeltMot & = & idle \\
DepositBeltIsReadyForLoading & = & false \\
PieceAtDepBeltEnd & = & true \\
CraneMagnet & = & on
\end{array}
$$

Figure 7.10: Liveness condition for initial states

With respect to the changes listed above and the given model of the environment no error could be found. The SMV tool returns the following output:

```
resources used:
user time:  10.1333 s, system time:  0.95 s
BDD nodes allocated:  56832
Bytes allocated:  18350080
BDD nodes representing transition relation:  25718 + 1
```

We conclude that the system model with respect to the given environment model satisfies the liveness property as required in the problem definition.

The Agent Progress Lemma Börger and Mearelli [BM97] introduce the *Agent Progress Lemma* in order to prove the liveness of the system. The proof of this lemma requires inspection of the rules. We suggest supporting the user with the model checker in order to check the interactive behaviour of the overall system.

The Agent Progress Lemma is split into six parts. Each part describes the progress of one of the agents with respect to the values of the interface functions. We formalise each of the parts in one CTL formula. The conjunct of these formulas gives the formalisation of the progress requirement of the agents.

Model checking the formulas of progress verifies that the SMV model satisfies the progress requirements if we choose the same initial condition as described in the last section. We conclude that the Agent Progress Lemma holds for the given system model with respect to the given environment model.

7.2.5   Discussion

The Production Cell model was chosen as a preliminary example of our approach of model checking ASM. It shows that in principle the SMV tool can be used to analyse ASM models. At the same time, however, this case study reveals a major drawback of this approach.

As an embedded system, the Production Cell involves a high degree of non-determinism due to the external functions that model sensor variables. We are not able to treat this kind of system properly by means of the SMV tool. Since additional assumptions on the behaviour of the environment (i.e., the sensor variables) must be modelled when doing model checking, we are forced to do so in a state-based way with a bounded number of choices. Such a state-based model, in contrast to a model based on temporal logic, restricts the unbounded non-determinism for possible behaviour. As a consequence, model checking decreases to a non-exhaustive test since not every possible behaviour is modelled.

However, the behaviour of the sensor variable *PieceInFeedBeltLightBarrier*, e.g., is nicely expressed in LTL by

$$\neg(FeedBeltFree) \Rightarrow \mathbf{F}\,(PieceInFeedBeltLightBarrier) \vee \mathbf{FG}(FeedBeltMot = \mathit{off})$$

This formula states that eventually the blank will arrive at the end of the belt or the feed belt motor will remain turned off. A similar model of the belt behaviour is not *finitely* expressible in a state-based way.

Using a deterministic environment or an environment model that is too restrictive may help to find some errors but it does not provide a reliable analysis method. In [Plo00], Plonka investigated the ASM model of the Production Cell by means of the model checker SVE ([F+96]). SVE is a model checker for CTL model checking that allows LTL formulas to be integrated into the system model. The tool is designed according to the work of Clarke, Grumberg, and Hamaguchi ([CGH97]) which is described in Section 3.4.2. Due to this extension, the environment model can be given in terms of LTL. Plonka found an error in the model of the Production Cell by means of increasing the choices of behaviour in the environment model. This error was not detected in our approach. This increasing of choices for the environment behaviour, however, may cause state explosion. For this reason, it was not possible to check liveness of the model with the approach given in [Plo00].

We conclude from this experience that a combined approach for CTL and LTL model checking would be helpful for analysing embedded systems. However, as the results in [Plo00] show, we need additional support for reducing complexity of the model checking task. This nicely motivates the work that is introduced in the next chapter: the interface from ASM to the MDG-Package.

Within the framework of MDGs, model checking algorithms are available that provide the same benefits as symbolic model checking for LTL discussed in Section 3.4: It is a symbolic model checking approach based on decision graphs and it allows the user to introduce logical assumptions into the system model. Furthermore, MDGs support abstraction. As the discussion above shows, it is crucial for most applications to reduce the complexity of the model's state space. Abstraction is a suitable means for this reduction.

Chapter 8

The MDG-Package

8.1 Interfacing the MDG-Package

In Chapter 6, the interface from the ASM Workbench (ASM-WB) to the SMV model checker is described. In this chapter, we focus on our second approach, the interface to the MDG-Package. This interface provides another means for analysing ASM models and supports abstraction.

The data structure that the MDG approach is based on are *multiway decision graphs* (MDGs) ([CZS+97], [CCL+97]). They are described in chapter 4 on page 51 as a generalisation of BDDs. The basic algorithms for MDGs (e.g., those described in Section 4.2.3) are prototypically implemented in Prolog by Zhou and others; [Zho96] is a recommendable documentation of this package.

In the literature, the MDG approach is introduced as a method for analysing hardware models given in a hardware description language called MDG-HDL. Invariant checking and reachability analysis is introduced in [CZS+97, CCL+97]. Applications of these approaches are described, for instance, in [ZST+96, LTZ+96]. Model checking of MDG-HDL specifications using a linear temporal logic is introduced in [Xu99, XCS+98]. MDG-HDL defines the front-end of the tool. To use it for model checking ASM, we would have to map ASM transition rules into MDG-HDL. However, this does not seem to be the closest mapping since any ASM model must be "simulated" (or re-modelled) in terms of a hardware description. Instead, it appears to be more reasonable to map ASM to MDGs directly. Doing so, we cannot use the checking framework of the MDG group as a black box tool with a language front-end as with SMV, for instance. Instead, we consider the underlying basic functionality of MDGs and tailor a checking tool for our own needs. That is, we provide an "interface" to the MDG-Package rather than to the "MDG-tool". This interface is realised by representing the ASM models through MDGs.

Our choice to extend our tool framework with an interface to the MDG-Package is mainly guided by three interests:

1. Since both notions, ASM and MDGs, are closely related to each other, the transformation from ASM into the MDG structure is easier and more concise than the treatment of the syntax of another input language (for instance, the SMV language) would be.

2. MDGs as a data structure for representing transition systems provide a powerful means for *abstraction* in order to fit large models into the model checking process.

3. Since the MDG-Package (cf. [Zho96]) is a well documented collection of basic functions that are sufficient to implement various types of model checking

algorithms, we are able to tailor a new checking tool according to our needs (in contrast to a black box tool for which either the code is not available or utterly undocumented). The basics for model checking $\mathcal{L}_{MDG}$ formulas[1] are given in [Xu99] for another input language. If we adapt these algorithms to our notation, the resulting tool will be an $\mathcal{L}_{MDG}$ model checker for ASM. This tool will nicely complement the facilities for checking CTL formulas given by the SMV tool (see the discussion in Section 9.2).

In the following, we go into detail about the three points for choosing MDGs as an appropriate data structure for supporting ASM analysis.

To start with the first claim, Section 8.2 compares ASM and MDGs. In Section 8.3, we describe our notion for supporting abstraction for ASM models based on MDGs. This should serve as a motivation for using MDGs for model checking ASM. We give an example in order to show how to use our approach. In Section 8.4, we introduce the transformation from ASM into MDG. Since we want to make use of abstract data that can be treated within the MDG approach, we have to extend the ASM language with syntax for abstract sorts. As a consequence, we have to extend the transformation algorithm that is described in Section 5.2. Subsection 8.4.1 introduces the extended transformation from ASM to ASM-IL. We can show that the extended transformation maps ASM models into an extended intermediate language ASM-IL$^+$ that satisfies the conditions for being a directed formula (see Section 4.2.1). These formulas can be represented by MDGs. Section 8.4.3 introduces the second step of the transformation that completes the interface from the ASM-WB to the MDG-Package.

8.2 ASM versus MDG

In order to justify the claim that is made in the first point above, we start with a comparison of the two notations. Although one is a specification language and the other one is a basic graph structure, they have a common base.

Many-sorted first-order logic. Both notations, ASM and MDGs, are based on many-sorted first-order logic: Gurevich uses the notion of many-sorted first-order structures to describe states of a system and adds transition rules for modelling the system behaviour during a run (see e.g., [Gur97] or Chapter 2). The MDG approach uses so called "*Abstract State Machines*" too in order to identify the system that is to be analysed. Its (sets of) states, transition and output relation are described by means of directed formulas (DF) which are a certain class of formulas of many-sorted first-order logic (see e.g., [CCL$^+$97]). This coincidence of the identical naming "Abstract State Machine" is understandable in that both constructs describe the same thing. However, Gurevich's ASM is a formal, high-level language for modelling where, in the MDG approach, the MDG-HDL language is used for modelling rather than their concept of abstract state machines.[2]

Abstract sorts. The difference between the languages, however, is given through the lack of *abstract sorts* in ASM language. To overcome this difference, we introduce this notion into ASM-SL as a kind of "syntactic sugar". Given this, we are able to indicate that a certain sort (or type) is abstract and thus every function that is applied to parameters of this type is either a cross-term or an abstract function

---

[1] $\mathcal{L}_{MDG}$ is a sub-language of Abstract_CTL* which is the universal fragment of first-order CTL* with abstract sorts.

[2] Note, that throughout this chapter we use the term ASM to denote the specification language and not the mathematical construct.

and has to be left *uninterpreted*. This distinct treatment of abstract sorts and the involved functions can be handled fully automatically and is implemented in our extended transformation algorithm.

ASM-IL$^+$ as MDG. The intermediate language for ASM (ASM-IL) that is introduced as a general interface language for the ASM-WB (see Section 5.2) is extended to a language ASM-IL$^+$ in order to treat abstract terms. ASM-IL$^+$ thus provides an interface to the MDG-Package. This format is similar to the notion of directed formulas (DFs) (see Section 4.2). DFs can be represented by MDGs in order to utilise the succinct representation and efficient algorithms defined on them (comparable to the OBDD algorithms that are used in other symbolic model checkers). The representation is canonic as shown in the literature (see [CZS$^+$97] for a proof).

The similarity of guarded updates in ASM-IL$^+$ and DFs is given through the fact that both describe simple state transition system. In ASM-IL$^+$, the transition relation is given as a set of tuples over locations and their guarded update set $\{loc_i', \{guard_{ij}, val_{ij} \mid 1 \leq j \leq m\} \mid 1 \leq i \leq n\}$. This set automatically provides a proper *partitioning of the transition relation* which helps to decrease the complexity of a model representation by decision diagrams (see discussion in Section 8.4.2). Moreover, the term unfolding (on concrete terms) that is implemented in the transformation algorithm ensures that all terms in our ASM-IL$^+$ model are concretely reduced. This condition is necessary to preserve the well-formedness of the resulting graph.

In Section 8.4.2, we show in more detail that an ASM model *after applying* our transformation (i.e., ASM-IL$^+$ models) preserves all of the well-formedness conditions on DFs and MDGs. Thus, ASM models can be represented by a set of well-formed MDGs.

In the next section, we discuss the support for abstraction that is provided through the MDG framework. Since model checking is introduced as a fully automatic tool support for analysis, it is important to provide support for the necessary abstraction task as well. We illustrate our notion of automatically generating abstract models once the data that should be abstracted are chosen. This is a very simple approach that still needs to be tested in practice.

8.3   Generating Abstraction

The MDG approach supports the distinction between concrete and *abstract sorts*. As a consequence various kinds of functions are relevant: concrete functions, abstract functions (with an abstract range), and so called cross-terms (functions with a concrete range but abstract parameters). Introducing the notion of abstract sorts and the facility for determining and distinguishing the kinds of functions provides a basis for automatically generating abstract models.

Applying abstraction to any given model is a useful means for reducing the size of the model. We strip off information that is not directly referred to in the properties to be checked. The abstracted terms and values can still be accessed within guards that influence the control flow of the system since any kind of boolean operator can be applied in order to compare their current evaluation in a state. These boolean operators yield a simple case distinction that can be explored exhaustively.

As a simple example, see the specification of a generic timer in Figure 8.1. The system gets as an input value any natural number *max* that specifies the number the timer has to count to. The timer has two states, COUNT and RING. As long as the system is in state COUNT it increments the state variable t in every step. Once t has reached the limit *max* the system changes into state RING; a bell might ring to give a signal. In the next step, t is reset and the timer starts again counting to *max*.

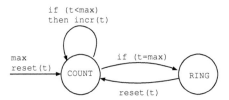

Figure 8.1: Example of a simple timer specification

This system specification can easily be abstracted by treating the natural numbers $\mathcal{N}$ as abstract sort $\mathcal{N}_{abs}$. We replace the equality relation on natural number by a new predicate *isEq* that compares two abstract values of sort $\mathcal{N}_{abs}$; $isEq(t, max)$ may evaluate to *true* if the (abstract) arguments are equal, otherwise it evaluates to *false*. The functions *incr* and *reset* turn into abstract functions that map any value of sort $\mathcal{N}_{abs}$ into a value of the same sort.

In the MDG approach, predicates like *isEq* are cross-term symbols; they are applied to abstract terms and evaluate to a value of concrete sort (in our example, boolean values). When model checking, we can use cross-terms since their range is a concrete type that can be enumerated. Without any knowledge of the value of the abstract parameter, we simply unfold the different (enumerate-able) cases and explore them when model checking. This may lead to exploring states that do not occur in the concrete model and hence may yield wrong counterexamples. We address this problem in the next section.

8.3.1   The Non-Termination Problem

Abstract functions and variables are treated as being *uninterpreted*, i.e., every interpretation is possible. We may describe the step of abstraction as stripping off semantics by giving up information about interpretation. Thus, the abstract specification is a model for all structures with the same signature. That is, an abstract model describes a set of concrete models with suitable (with respect to the signature) interpretations for abstract sorts, functions, variables, and cross-terms. Figure 8.2 depicts the abstraction step: We consider the sort Q in our concrete model as abstract and change all its occurrences into $Data_{abs}$. As a result, we get an abstract model of the same signature. For this abstract specification, all those interpretations are suitable that have a sort, a 2-ary function that maps arguments of the given sort to a value of the same sort, and a boolean predicate over the sort. In the figure, we give some examples for different interpretations for the sort $Data_{abs}$ and the functions that are possible.

The purpose of this abstraction step is to substitute infinite sorts, and functions over them, since these cannot be exhaustively explored. The use of cross-terms on abstract sorts and their complete case distinction naturally provides a partitioning of the infinite sorts into finitely many equivalence classes. The state space of the abstract model is smaller in most cases. It can be canonically represented by MDGs and the corresponding model checking algorithm can be applied to check certain properties.

In some cases, however, we run into the problem of *non-termination* or *false negatives* (see [CZS+97], [CCL+97], [MSC97]). The abstract model may allow an interpretation that includes cases that are excluded for the concrete model that we

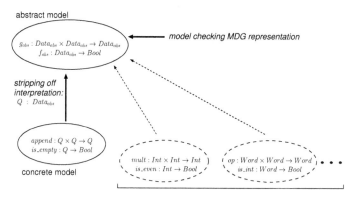

Figure 8.2: Lifting a concrete model to an abstract model

want to investigate. For instance, in the concrete model over the sort of queues, denoted as Q in Figure 8.2, it is obviously the case that if we append two queues that are not both empty the resulting queue is not empty either:

$$(\neg(is\_empty(q_1)) \vee \neg(is\_empty(q_2)) \;\Rightarrow\; \neg(is\_empty(append(q_1, q_2))).$$

However, in another interpretation, e.g., over integers Int, the same rule might be violated:

$$(\neg(is\_even(q_1)) \vee \neg(is\_even(q_2)) \;\not\Rightarrow\; \neg(is\_even(mult(q_1, q_2))).$$

If the value of the predicate *is_empty* is determining the control flow, the abstract model may include behaviour that is not intended in the original model. This unintended behaviour may cause *false negatives* (we called them wrong counterexamples in Section 5.1) or may lead to a reachability analysis that does not terminate although the intended concrete model satisfies the termination criteria.

One means to overcome the problem caused by unintended interpretations is to add axioms over the abstract functions and cross-terms of our (partly abstract) signature. Adding axioms to the abstract model can be seen as a refinement step: some information that was lost because of the abstraction step is introduced into the model again (see Figure 8.3). These axioms can be formulated as rewrite-rules (denoted as `rr(...)` in the figure) which are fed into the Prolog system that computes the MDGs and runs the model checking procedures. Basically, rewriting in Prolog causes a substitution of any term that matches the pattern in

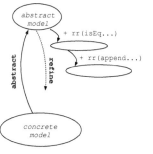

Figure 8.3: Adding axioms for refinement

the rewrite rule. Adding rewrite rules into the system is supported by a function that is available in the MDG-Package (see [Zho96]).

Obviously, it has to be proved that the rewrite rules that are added to the abstract model are correct for the concrete model we want to check. For instance, in the example in Figure 8.2, we have to proof that the axiom (to be modelled in terms of the abstract model)

$$\forall q_1, q_2 \in Data_{abs} \; : \; (\neg(f_{abs}(q_1)) \vee \neg(f_{abs}(q_2))) \; \Rightarrow \; \neg(f_{abs}(g_{abs}(q_1, q_2)))$$

is satisfied for the interpretation ψ that maps the abstract sort and function symbols to the sort and functions of the concrete model, i.e., $\psi(Data_{abs}) = Q$, $\psi(f_{abs}) = is\_empty$ and $\psi(g_{abs}) = append$.

If the system developer who uses the MDG approach fails to add all rewrite rules that are necessary for avoiding the non-termination problem, it is still guaranteed that the model checking algorithm is *correct* since it never gives a false *positive* result. That is, if the result of the checking procedure states that the model satisfies the properties to be checked then this holds for *any* interpretation of our model. In the worst case, the MDG checking is either not applicable because it does not terminate, or it yields a counterexample that is not a proper run for the concrete system at hand. In the latter case, the user will identify this as a false negative when investigating its interpretation in the concrete model.

Other approaches for generating an abstraction of a model prove that the chosen abstraction function is *property preserving* (see Section 9.1). This proof would be comparable with a proof that the set of added rewrite rules *exhaustively* reflects the properties on abstract functions and cross-terms and, therefore, no false negatives could occur. This is a much more complex proof task than the one that is intended in our approach (c.f., [SS99, LGS+95]).

8.3.2   Automatic Support for the Generation

Lifting a model to a higher level of abstraction can be done automatically within our transformation step from ASM to MDG provided the developer has chosen the domains of the ASM that should be considered as abstract sorts. Functions in the ASM model that involve abstract data are automatically treated as abstract functions or cross-terms. In contrast to any other approach of generating abstraction, there is no extra effort necessary to change the model under investigation other than the change of the data type definition.

- Any function symbol that maps into an abstract sort is treated as an *abstract function* (e.g., function f_{abs} in Figure 8.2). These functions are left uninterpreted.

- Any function that applies to some arguments of abstract sort but maps into a concrete sort are treated as *cross-terms* (e.g., function g_{abs} in Figure 8.2).

Abstract functions are not unfolded by our transformation algorithm. We say that abstract functions are left *uninterpreted* since unfolding involves the evaluation of a term (see Section 5.2 and in particular Figure 5.4 on page 71). Obviously, abstraction simplifies the transformation (less terms have to be unfolded) and also reduces the size of the resulting model to be checked (unfolding of terms causes the generation of many rules instances). (See Section 8.4.1 for more details of our extended transformation that can handle abstract sorts and functions.)

Operators that are built-in in the target language, the *primary operators* (see Section 5.2.1 on page 75), are not unfolded in our transformation algorithm, i.e., we keep the corresponding terms as they are. These operators have to be renamed when they become abstract in order to avoid name clashing between the abstract and the concrete function symbol. In particular, boolean predicates over a domain that might be abstracted (e.g., numbers) are renamed in order to denote the "lifted" cross-term. The following renaming mapping for predicates is implemented in our transformation algorithm and thus automatically applied during the transformation:

$$
\begin{aligned}
(a = b) &\Rightarrow isEq(a, b) \\
(a \neq b) &\Rightarrow isNeq(a, b) \\
(a < b) &\Rightarrow le(a, b) \\
(a <= b) &\Rightarrow leq(a, b) \\
(a > b) &\Rightarrow ge(a, b) \\
(a >= b) &\Rightarrow geq(a, b)
\end{aligned}
\qquad (8.1)
$$

To give an example, we consider the term $x = M$ in a guard. In the original model both terms x and M are of some concrete but possibly infinite type. We may change the model by editing the sorts of x and M to be abstract. Correspondingly, the equation will be transformed into the cross-term $isEq(x, M)$. To complete the expression for building an MDG (c.f., the well-formedness conditions on MDG in Section 4.2), we add the concrete equality: $isEq(x, M) = true$.

For the most common boolean predicates, new cross-operators are pre-defined as shown in the mapping in (8.1). These cross-operators provide a natural partitioning on concrete sorts such as integers, real numbers, ordered sets, etc. The mapping may be extended according to a particular problem domain that involves other predicates. Any other function symbol that is applied or mapped to a term that is turned into an abstract term is left unchanged.

Our transformation (see Section 8.4.1) is a useful means for supporting abstraction of models that are not treatable by the checking algorithm. The generation of the abstract model is done automatically according to the different treatment of concrete and abstract functions or cross-terms. New function symbols for boolean functions that turn into cross-terms are introduced automatically. In the following section, we give a simple example of abstracting an ASM model.

8.3.3 A Simple Example

We use the simple example of the timer given in the last subsection in order to illustrate how to generate an abstract model. Figure 8.4 shows a *concrete* model of the timer that counts to some number `max` which is given as input.

The transition rules of our timer model reflect the behaviour of the state machine in Figure 8.1 on page 118. We can check this model for correctness for a particular input value (max $< n$), e.g., for an input 10 the timer should never exceed this limit.

Assume that we want to check the correct behaviour of the timer not only for any maximum that is less than a particular n but for an arbitrary n. We can use abstraction to do this. As described in the last subsection, the abstraction step affects only the definition of the domains and the declaration of functions, i.e., the state space definition. Therefore, Figure 8.5 which depicts the abstract model of the timer, only shows the state space definition; the transition rules remain unchanged. Note that in the given example the benefit of our abstraction support is less obvious since the model consists of only four rules, which are very simple. Other models, however, may have a large amount of very complex rules. The effort required to remodel these rules by hand can be significant.

When abstracting a concrete model, we define some domains as being abstract. In Figure 8.5 the domain `TIME` turns into an abstract domain. The notation for an abstract sort is a singleton that contains a string that starts with the substring "abstract". This is a workaround for we lack a proper keyword in ASM-SL, the input language of the ASM-WB. However, it still allows using the internal type-checker

---

$\texttt{MODE} = \{\,count,\,ring\,\}$
$\texttt{TIME} = \{1..n\}$

external function **max** : TIME
static function   **reset** : TIME $\rightarrow$ TIME $\;==\;$ $\{t \mapsto 0 \mid t \in$ TIME$\}$
static function   **incr** : TIME $\rightarrow$ TIME $\;==\;$ $\{t \mapsto (t+1) \mid t \in$ TIME$\}$
dynamic function **mode** : MODE
dynamic function **timer** : TIME

    if (mode = *count*) $\wedge$ (timer < max)
      then timer := incr(timer)

    if (mode = *count*) $\wedge$ (timer = max)
      then mode := *ring*

    if (mode = *ring*)
      then mode := *count*
           timer := *reset*(timer)

Figure 8.4: The concrete ASM model for the timer

of the ASM-WB[3]. Since the type-system of ASM-SL allows only a strongly typed signature (i.e., each term has a unique sort), the strings that indicate abstract sorts must differ in the substring that follows the prefix "abstract". In our automatic treatment of abstracted models, we recognise an abstract sort by matching only the initial substring of this string. This way, we can introduce more than one abstract sort.

Static functions that turn into abstract functions map an abstract value to an abstract value. For example, the functions **reset** and **incr** are defined in Figure 8.5 by the mapping *abstractTime* $\rightarrow$ *abstractTime*. These functions are not evaluated in our transformation since they are of abstract sort.

Applying the extended transformation algorithm the ASM model is simplified and flattened to a set of guarded updates for every primed location. We get a list of tuples of the form $\{loc_i, [(guard_{ij},\, val_{ij})]\}$:

$$\{(\textbf{timer'}, [(((\text{mode} = count) \wedge (le(\text{timer, max}) = true)), \text{incr}(\text{timer})),$$
$$((\text{mode} = ring), \text{reset}(\text{timer}))\;])$$
$$(\textbf{mode'}, [(((\text{mode} = count) \wedge (isEq(\text{timer, max}) = true)), ring),$$
$$((\text{mode} = ring), count)\;])\}$$

Note that the "values" in the guarded update pairs need not be values anymore. We may have abstract functions as well (**incr(timer)** and **reset(timer)** in our example). These guarded updates can be represented in an MDG as shown later.

In order to avoid non-termination or even to reduce the state space of the model, axioms can be added by means of rewrite rules. These additional rewrite rules provide a means for refining the abstract model successively (c.f., Figure 8.3). In

---

[3]Since we do not evaluate abstract terms, this notation for abstract sorts (that models them as being enumerated rather than infinite) has no impact on our approach.

MODE $= \{\, count, ring\, \}$
TIME $= \{\, abstractTime\, \}$

external function **max** : TIME
static function   **reset** : TIME $\rightarrow$ TIME $==$ $\{\, abstractTime \mapsto abstractTime\, \}$
static function   **incr** : TIME $\rightarrow$ TIME $==$ $\{\, abstractTime \mapsto abstractTime\, \}$
dynamic function **mode** : MODE
dynamic function **timer** : TIME

Figure 8.5: The abstract ASM model for the timer: the state space

the given example, we add rewrite rules over the cross-terms $isEq()$ and $le()$ in order to exclude not-intended interpretations of the function symbols. For instance $\forall a, b$:

$$
\begin{aligned}
isEq(a, b) &\Rightarrow \neg(le(a, b)) \\
le(a, b) &\Rightarrow \neg(isEq(a, b)) \\
le(a, b) &\Rightarrow \neg(le(b, a)) \\
isEq(a, b) &\Rightarrow isEq(b, a)
\end{aligned}
$$

These rewrite rules can be expressed as rewrite rules in the MDG-Package. A rewriting function removes every path in the reachability graph that contains a contradictory conjunct. That is, no path will comprise node-edge pairs of the form $(isEq(a, b), true)$ and $(le(a, b), true)$ at the same time. Accordingly, the graph that represents the reachable states becomes smaller. For a better support, we add a set of ordinary rewrite rules over equality, $<$, $\leq$, $>$, and $\geq$ automatically during our transformation.

In the next section, we introduce the adaption of the transformation algorithm in order to treat abstract sorts and functions as we suggested above. In a second step, the guarded updates of the extended ASM interface language (ASM-IL$^+$) are transformed into MDGs. This completes the interface from the ASM-WB to the MDG-Package.

8.4 The Transformation of ASM into MDG

In order to benefit from the notion of abstract sorts for supporting abstraction, we have to extend the ASM language with a syntactic feature that allows a sort to be marked as abstract. This extension requires an extension of the basic transformation algorithm as well: We have to adapt the first step of the transformation (simplifying and unfolding of ASM terms and rules, see Section 5.2 on page 68) to the treatment of abstract functions. These functions are left uninterpreted.

This section is split into two parts: Firstly, we detail the adaption of the basic transformation algorithm that transforms ASM models into the extended intermediate language ASM-IL$^+$. Secondly, we develop a transformation from ASM-IL$^+$ models into MDGs. We have to justify that ASM-IL$^+$ rules represent directed formulas of a many-sorted first-order logic (DF) (see Section 4.2 on page 55). These formulas can be canonically represented by MDGs. The second transformation step completes the interface from the ASM Workbench to the MDG-Package.

8.4.1 The Adapted Transformation into ASM-IL$^+$

The original transformation algorithm maps all occurrences of dynamic and external functional terms $f(t_1, \ldots, t_n)$ into locations $f(a_1, \ldots, a_n)$, which can be renamed to simple state variables. This is done by simplifying and unfolding: The evaluation of parameter terms t_i to elements a_i may be state dependent. Each possible evaluation of these terms leads to a different location. The result of the unfolding procedure is a set of locations that are given by all possible evaluations of the dynamic parameter terms.

Functions and terms of abstract sort cannot be similarly evaluated since their interpretation is not specified. Unfolding a term involves evaluating it, i.e., adding the state dependent interpretation to a term. Therefore, an *uninterpreted* function cannot be unfolded in any term in which it appears. In order to implement this special treatment for abstract functions, we have to restrict the unfolding procedure to functions that are concrete.

Adapted simplification function. The transformation algorithm for simplification and unfolding follows an inductive schema. The term simplification $[\![.]\!]_\zeta$ is defined in Section 5.2 on page 68. The treatment of abstract functions and cross-terms is easily introduced into this inductive schema by adding some case distinctions to the term simplification: abstract terms should not be unfolded to any evaluation. The modified definition for term simplification is summarised in Table 8.1.

The simplification function $[\![.]\!]_\zeta$ is related to the variable assignment[4] $\zeta : V \rightarrow S^A$ since all concrete variables in V are mapped to the values of the corresponding concrete sort.

The base of the induction schema is given by simple terms. We distinguish constant *values* a, *locations* loc, that may change their values at a transition step, and *variables* v, which can be evaluated to values a if they are **concrete** (according to ζ) or left as uninterpreted variables v^{abs} if they are **abstract**[5]. Applying the unfolding function to simple terms changes the term only in the case of a concrete variable that can be evaluated. Constants, locations, and abstract variables remain unchanged.

We extend this base by *functional terms*: Functional terms are **abstract** functions applied to terms t_i that can be abstract terms or concrete values. In Table

---

[4] Recall that S^A denotes the universe (or carrier set) of the sort S in state A.

[5] The index *abs* is used in Table 8.1 for indicating that a variable or term is of abstract sort or is a cross-term.

Adapted Term Unfolding

for simple terms we distinguish

$$[\![a]\!]_\varsigma = a \ \ \text{values} \quad \text{and} \quad [\![loc]\!]_\varsigma = loc \ \ \text{locations}$$

for variables we get

$$[\![v]\!]_\varsigma = \begin{cases} a = \varsigma(v) & \text{if } v \in \text{dom}(\varsigma) \text{ and of \textbf{concrete} sort} \\ v^{abs} & \text{otherwise} \end{cases}$$

for applied functions we distinguish

$$[\![t_i]\!]_\varsigma = a_i \quad \text{for each } i \in \{1, \ldots, n\} \ \Rightarrow$$

$$[\![f(t_1, \ldots, t_n)]\!]_\varsigma = \begin{cases} a = f^A(a_1, \ldots, a_n) & \text{if } f \text{ is a static} \\ & \text{function name of \textbf{concrete} sort} \\ loc = (f, (a_1, \ldots, a_n)) & \text{if } f \text{ is a dynamic/external} \\ & \text{function name of \textbf{concrete} sort} \\ f^{abs}(a_1, \ldots, a_n) & \text{if } f \text{ is a static} \\ & \text{function name of \textbf{abstract} sort} \\ loc^{abs} = & \text{if } f \text{ is a dynamic/external} \\ (f^{abs}, (a_1, \ldots, a_n)) & \text{function name of \textbf{abstract} sort} \end{cases}$$

if all arguments are of **constant** sort and
$$[\![t_i]\!]_\varsigma = loc \quad \text{or} \quad [\![t_i]\!]_\varsigma = f'(\bar{t}') \quad \text{for some } i \in \{1, \ldots, n\} \ \Rightarrow$$
$$[\![f(t_1, \ldots, t_n)]\!]_\varsigma = f([\![t_1]\!]_\varsigma, \ldots, [\![t_n]\!]_\varsigma)$$

if some arguments t_i are of **abstract** sort and $[\![t_i]\!]_\varsigma = t^{abs} \ \Rightarrow$

if f is a **cross-term** operator or a **static** function name:
$$[\![f(t_1, \ldots, t_n)]\!]_\varsigma = f([\![t_1]\!]_\varsigma, \ldots, [\![t_n]\!]_\varsigma)$$

if f is an **abstract** dynamic/external function name:
$$[\![f(t_1, \ldots, t_n)]\!]_\varsigma = (f^{abs}, ([\![t_1]\!]_\varsigma, \ldots, [\![t_n]\!]_\varsigma))$$

Table 8.1: Adapted Term Transformations

8.1, we denote these abstract functional terms as $f^{abs}(t_1, \ldots, t_n)$. They cannot be unfolded further and the simplification terminates.

Any dynamic or external function turns into a *locational term*. Locational terms can be simple, i.e., concrete locations loc, abstract functional terms loc^{abs}, or cross-terms $f(t_1, \ldots, t_n)$. External cross-terms can be fully explored since their range is finite and thus can be enumerated.

Locational terms that are abstract functions (and not cross-terms) are mapped into simple variables by merging function symbol and parameter symbols into one new variable name. This is indicated through the pair $(f^{abs}, (t_1, \ldots, t_n))$. In contrast, any static functional term, and also any cross-term, is kept as a function application: $f(t_1, \ldots, t_n)$ (c.f., the notation in Table 8.1).

The term simplification $[\![.]\!]_\varsigma$ (shown in Table 8.1) works inductively over arbitrary function application $f(t_1, \ldots, t_n)$. We distinguish between terms whose parameters are all concrete values, terms whose parameters are concrete but have to be unfolded further, and terms that have some parameters of abstract sort:

1. If all parameters are (unfolded to) values, we distinguish between

    - static functions that are **concrete**; these have to be evaluated by means of applying the function interpretation f^A (which is the same in any state A) to the parameter values;

    - dynamic or external functions that are **concrete**; these turn into a location *loc* by means of merging function symbol and parameter symbols; any location *loc* is thus equivalent to a simple state variable;

    - static functions of **abstract** sort; these are left as uninterpreted functions f^{abs} which are not applied; they are denoted as functional terms $f^{abs}(a_1, \ldots a_n)$.

    - dynamic or external functions of **abstract** sort; we denote these terms as loc^{abs}, a locational term that can change its value, and proceed similar to concrete locations: we merge the function symbol and the parameter symbols and get an abstract state variable.

2. If all parameters are of **concrete** sort but some of them are non-values, i.e., locations *loc* or other function applications $f'(\bar{t}')$ that have to be unfolded further, we apply the unfolding function to the parameters first.

3. If some parameters are **abstract** terms t^{abs}, i.e., abstract variables v^{abs} or abstract functions $f^{abs}(t_1, \ldots, t_n)$, we simplify the parameters by $[\![.]\!]_\varsigma$ first; two cases may occur:

    - the function is a cross-operator or an abstract static function; in this case we keep the function;

    - the function is an abstract dynamic/external function; we merge the function symbol and the parameter symbols into one abstract variable.

Adapted Term Unfolding (cont.)

quantified terms over finite ranges
$$\{a \in S_i^A \mid B \models g(a)\} \text{ where } S_i^A \text{ is a } \textbf{concrete} \text{ sort:}$$

$$[\![((\exists v : g(v)) \ s(v))]\!]_\varsigma \; = \; [\![s(v)]\!]_{\varsigma[v \mapsto a_1]} \vee \ldots \vee [\![s(v)]\!]_{\varsigma[v \mapsto a_n]}$$

$$[\![((\forall v : g(v)) \ s(v))]\!]_\varsigma \; = \; [\![s(v)]\!]_{\varsigma[v \mapsto a_1]} \wedge \ldots \wedge [\![s(v)]\!]_{\varsigma[v \mapsto a_n]}$$

quantified terms over **abstract** ranges
$$\{a \in S_i^A \mid B \models g(a)\} \text{ where } S_i^A \text{ is an } \textbf{abstract} \text{ sort:}$$

$$[\![((\forall v : g(v)) \ s(v))]\!]_\varsigma \; = \; [\![s(v)]\!]_\varsigma$$

($[\![((\exists v : g(v)) \ s(v))]\!]_\varsigma$ is not defined; see remark below)

Table 8.2: Adapted Term Transformation of first-order Terms

First-order terms of the form $(\exists v : g(v)) \ s(v)$ or $(\forall v : g(v)) \ s(v)$ are simplified as in the original transformation as long as their range is a concrete and finite set $\{a \in S_i^A \mid B \models g(a)\}$ (see Definition (2.10)). If the range of a first-order term turns into an abstract range by means of applying abstraction to the domain S_i^A

then the range is not enumeratable any more. The simplification keeps the head variable v as an abstract variable and treats the body $s(v)$ as a cross-term. This is shown in Table 8.2 by applying $[\![.]\!]_\varsigma$ to the body. For abstract head variables v, $[\![s(v)]\!]_\varsigma$ implicitly models universal quantification over every $a \in S_i^A$. Existential quantification over an abstract variable is logically not expressible.

Note that uninterpreted, abstract functional terms $f^{abs}(t_1, \ldots t_n)$, where f^{abs} is an abstract static function and t_i is a value or another abstract term, or $f(t_1, \ldots t_n)$ where f is a cross-term operator, are not mapped into a state variable name. In contrast to the SMV approach, these functional terms may appear as labels in the MDG structure. It is easy to see that abstraction saves alot of unfolding effort. The resulting set of guarded updates is much smaller.

Rule Simplification

$[\![\texttt{skip}]\!]_\varsigma \;=\; \texttt{skip}$

$[\![f(\bar{t}) := t]\!]_\varsigma \;=\; \texttt{if } \mathit{true} \texttt{ then } [\![f(\bar{t})]\!]_\varsigma \; := \; [\![t]\!]_\varsigma$

$[\![\texttt{block } R_1 \ldots R_n \texttt{ endblock}]\!]_\varsigma \;=\; [\![R_1]\!]_\varsigma \ldots [\![R_n]\!]_\varsigma$

$[\![\texttt{if } g \texttt{ then } R_1 \texttt{ else } R_2]\!]_\varsigma \;=\; \begin{cases} \texttt{if } [\![g]\!]_\varsigma \texttt{ then } [\![R_1]\!]_\varsigma \\ \texttt{if } \neg [\![g]\!]_\varsigma \texttt{ then } [\![R_2]\!]_\varsigma \end{cases}$

Table 8.3: Rule Simplifications

Table 8.3 summarises the rule simplification that applies the simplification function $[\![.]\!]_\varsigma$ (c.f., Table 8.1) to the terms in the rules. This rule simplification is not changed. We recall the definition here for the readers' convenience.

Adapted rule unfolding. Table 8.4 defines the extended rule unfolding. In this rule unfolding, simple update rules are left unchanged. More complex rules are unfolded according to the evaluation of **concrete** locations that can be found in the rule. For every possible evaluation a_i we introduce an instantiation of the rule $R[loc/a_i]$ that is guarded by the equation $(loc = a_i)$, where loc is the first concrete location occurring in the rule R. The substitution $R[loc/a_i]$ formalises that every occurrence of loc is substituted by the value a_i. Note that **abstract** locations are not unfolded; they are left uninterpreted.

Also, all locations that appear as a left-hand side (LHS) of an update or that are parameters of the *primary* operators (e.g., equality, conjunction, disjunction, and negation) are excluded from the unfolding procedure. This is due to optimisation issues and already discussed in Section 5.2.1 on page 75.[6]

ASM-IL$^+$ representation. As a result of the adapted term simplification and rule unfolding we get a list of pairs over locations and their guarded updates, which is the ASM-IL$^+$ representation for the ASM model. In contrast to ASM-IL representations of an ASM model, these pairs may contain abstract variables and cross-terms as LHSs and RHSs, as well as abstract functions as RHSs of equations or updates. To summarise the result of the adapted transformation algorithm, we give the following definition.

---

[6]Note that in the MDG approach, we do not consider arithmetic operations as being primary.

Adapted Rule Unfolding

If $R = loc_1 := a_1 \ldots loc_n := a_n$:

$$\mathcal{E}(R) = R$$

Otherwise:

$$\mathcal{E}(R) = \begin{cases} \texttt{if } loc = a_1 \texttt{ then } \mathcal{E}(\llbracket R[loc/a_1] \rrbracket_\varsigma) \\ \ldots \\ \texttt{if } loc = a_n \texttt{ then } \mathcal{E}(\llbracket R[loc/a_n] \rrbracket_\varsigma) \end{cases}$$

where
loc is the first location **of concrete sort** occurring in R but not as an LHS of an update rule and not as a parameter of a primary operation; $\{a_1, \ldots, a_n\}$ is the range of location loc.

Table 8.4: Unfolding of concrete Locations in a Rule

Definition 8.23:   Location-update pair

A *location-update pair* is a tuple of the following form:

$$(\ loc\_term_i, \ [\ (guard_{i1}, upd\_term_{i1}), (guard_{i2}, upd\_term_{i2}), \ldots]\ )$$

A locational term loc_term_i can be either a concrete location loc, an abstract location loc^{abs} or a dynamic/external cross-term $f(\bar{t})$. Any concrete parameter is unfolded and the abstract parameters are kept.

Any guard $guard_i$ may consist of conjunction, disjunction and equation as primary (not simplified) operations, where operands can be (unprimed) locational terms, variables and abstract functions (these only as a RHS of an equation).

The right-hand side (RHS) of an update upd_term_{ij} can be a concrete value a, a location loc, an abstract functional term $f^{abs}(\bar{t})$, or a cross-term $f(\bar{t})$.

We call a pair $(guard_{ij}, upd\_term_{ij})$ in a location-update pair a *guarded-update pair*.

Limitations of the adapted term simplification

We identify two cases in which the adapted term simplification does not provide proper results and thus is not applicable:

1. Any dynamic or external abstract function is mapped into an abstract state variable. For any concrete parameter (that can be unfolded) we create several instances of this state variable. However, abstract parameters are not unfolded; instead of multiple instances we get only one state variable. For example, assume that in the concrete ASM we have a sort $A = \{a, b, c\}$ and a dynamic function $f : A \rightarrow B$. The location $f(x)$ is simplified into three state variables $\mathtt{f\_a}$, $\mathtt{f\_b}$, and $\mathtt{f\_c}$. Abstracting this model we may change the sorts A and B into abstract sorts. The same location $f(x)$ is now simplified into one variable $\mathtt{f\_x}$. The resulting abstract ASM-IL$^+$ model is not a correct abstraction of the concrete model. Note that the same problem does not occur for cross-terms: any function $g : A \rightarrow C$, where C is not an abstract sort, is kept as a function application rather than being mapped into a state variable.

2. Existentially quantified first-order terms $(\exists v : g(v))\ s(v)$ where the sort of variable v turns into an abstract sort cannot be simplified properly. The existence of a witness (necessary to evaluate existential quantification) is not decidable since an abstract sort has no concrete (interpreted) entities. A similar problem does not occur for universally quantified terms.

As a consequence of these limitations, we get guidance as to where abstraction should not be applied in our approach: Any sort that is used as a domain (or part of a cross-product of domains) of an n-ary dynamic or external function should not be abstracted. For example, we may not abstract the set of agents in a multi-agent ASM since a variable *Self*, ranging over agents, is used as parameter for any function that is local to an agent. Also ranges of head variables in first-order terms should be left enumerable. These restrictions, however, appear reasonable.

Correctness

Since the adapted transformation algorithm is identical to the original transformation algorithm for terms without abstract sorts, its correctness is based on the proof in Section 5.2. In the case of terms with abstract sorts, the adapted transformation algorithm itself defines the semantics of these terms and therefore no correctness proof is required.

A set of location-update pairs represents an *abstract* transition system, which can be treated by a tool that is based on MDGs. In the next subsection, we show that each location-update pair represents a *directed formula* (DF) in terms of the MDG approach.

8.4.2   ASM-IL$^+$ Models as Directed Formulas

When we apply the simplification and unfolding algorithms to ASM models with abstract sorts and functions, we get an ASM-IL$^+$ representation: the location-update pairs. If we want to map a set of location-update pairs into MDGs, we have to justify that the well-formedness conditions of MDGs are satisfied. In this subsection, we show that our transformation provides directed formulas (DFs) that can be canonically represented by well-formed MDGs (we follow the description of DFs and MDGs in Section 4.2 on page 55).

Concretely reduced terms. Concretely reduced DFs can be represented by well-formed MDGs. All concrete terms on the RHS of an update or equation in such formulas are individual constants. As introduced in the last section, the simplification function $[\![.]\!]_\varsigma$ unfolds all concrete terms that appear in ASM rules. This way, all terms in location-update pairs are already *concretely reduced* in the sense of the well-formedness conditions of MDGs (c.f., Section 4.2 on page 55), i.e., all concrete functions and variables are mapped into their values.

Partitioned transition relation. In model checking approaches that are based on decision diagrams (e.g., BDDs), the transition relation of a transition system that is to be checked should be partitioned and represented by smaller graphs rather than represented and treated as one large graph. The notion of a *partitioned transition relation* is introduced in [BCMD90] and [BCL91] (see also [BCL+94]). For graph-based approaches, partitioning helps prevent the (single) representing graph growing too big. Instead of working with one graph for representing the whole transition relation, the algorithms work on a set of smaller graphs, each representing a part of the transition relation only. This technique is adapted for the MDG approach as well (see [ZSC⁺95]). All algorithms (e.g., relational_product, pruning_by_subsumption, etc.) expect lists of MDGs as input to operate on.

Since in ASM-IL⁺ every location has an attached list of guarded-update pairs, this representation naturally provides a partitioning of the overall transition relation. Moreover, our partitioning naturally yields MDGs with disjoint sets of primary abstract variables (see below).

Correctness of the mapping

The correctness of mapping location-updates pairs into MDGs is based on the following lemma (recall that MDGs are a canonical representation for directed formulas).

Lemma 8.24:
Each location-update pair $(loc\_term_i, [(guard_{i1}, upd\_term_{i1}), \ldots])$ represents a directed formula of many-sorted first-order logic.

Proof: We can prove Lemma 8.24 by relating the well-formedness conditions that are given for DFs (see Section 4.2) to the location-update pairs that may occur in an ASM-IL⁺ representation. As introduced in [CCL⁺97], a directed formula is a formula in disjunctive normal form (DNF) such that each disjunct is a conjunct of equations. A similar formula is given by our ASM-IL⁺ pairs:

$$(loc\_term_i, [(guard_{i1}, upd\_term_{i1}), \ldots]) \Leftrightarrow$$
$$\bigvee_j (loc\_term'_i = upd\_term_{ij} \land guard_{ij}) \qquad (8.2)$$
$$\lor \left(loc\_term'_i = loc\_term_i \land \bigwedge_j \neg guard_{ij} \right)$$

where $loc\_term'_i$ denotes the locational term loc_term_i in the next state. The first part of the disjunction conjoins all guards and corresponding updates. The second part specifies the "else-case", i.e., if none of the guards are true then the location should keep its old value. According to the adapted transformation algorithm that is introduced in the last subsection, this DNF may contain the following:

- $loc\_term'_i$ is

    - a location of **concrete** sort $loc = (f, (a_1, \ldots, a_n))$ which we identify with a state variable: any tuple of the form $(f, (a_1, a_2 \ldots, a_n))$ will be merged into a state variable identifier $f\_a_1\_a_2\_\ldots\_a_n$;

    - a locational term of **abstract** sort $loc^{abs} = (f^{abs}, (t_1, \ldots, t_n))$, where t_i can be a value a, an abstract variable v^{abs} or a concrete/abstract locational term $(h, (t'_1, \ldots, t'_m))$; function symbol and parameter symbols are merged into one state variable name similar to concrete locations;

    - or a **cross-term** $g(t_1, \ldots, t_n)$ which is treated as a functional term rather than a variable; the cross-operator and the parameters are *not* merged into one name; each parameter t_i can be a value a, an abstract variable v^{abs} or a concrete/abstract locational term $(h, (t'_1, \ldots, t'_m))$;

- upd_term_i is
  - a value a, i.e., an individual constant of concrete sort
  - a concrete or abstract (unprimed) location loc_term (this includes concretely reduced cross-terms);
  - an abstract variable v^{abs}
  - a term $g(t_1, \ldots, t_n)$ of abstract sort containing no (primed) next state variables.

- $guard_i$ is a boolean expression over equations that do not contain primed locational terms $loc\_term'_i$. Their LHS and RHS can be a concrete location or value, or a cross-term. Any equation with an abstract variable or function, e.g., $a = b$, is transformed into an equation involving a new cross-term, e.g., $isEq(a, b) = true$ (for more details see Section 8.3.2 on page 120).

- The only abstract locational term that may appear as an LHS of an equation is $loc\_term'_i$. At the same time this abstract location appears in every disjunct of the DNF in (8.2).

- Any conjunct of equations with the same LHS is simplified during the transformation; equations appear only among the guards in location-update pairs since the only other LHS is the primed locational term (which does not occur in the guards); according to the rule simplification given in Table 8.3 every guard is simplified by $[\![.]\!]_\varsigma$. As a consequence, all remaining LHSs in a disjunct are pairwise distinct.

Given this, any location-update pair is a DF of type $U \to V$ according to the definition in Section 4.2. The set of "independent" variables U is the set of unprimed locational terms Loc of concrete or abstract sort, including the cross-operators that are dynamic. These occur in the guards and thus characterise the current state in which the update can be fired. Also, the set of abstract variables V^{abs} belongs to the set of independent variables U. Abstract variables may occur as not-unfolded parameters in (universally) quantified terms. That is $U = Loc \cup V^{abs}$.

The set of "dependent" variables V is the set of primed locational terms which is a singleton for each location-update pair: $Loc' = \{loc\_term'_i\}$. It defines the transitions to possible next states if any of the updates fire. Thus, every location-update pair is a DF of type $(Loc \cup V^{abs}) \to Loc'$.   ∎

The above has shown that location-update pairs are DFs. Each of these DFs has at most one primary abstract variable: $loc\_term'_i$. As a consequence each location-update pair can be represented as an MDG. These MDGs have pairwise disjoint sets of primary abstract variables. This property is a necessary precondition for applying the algorithms for computing the relational product or the conjunction (see Section 4.2). In the next section, we introduce our algorithm that interfaces ASM-IL$^+$ with the MDG-Package by means of generating MDGs from the location-update pairs.

8.4.3   Transformation of ASM-IL⁺ into MDGs

In the second step of our transformation, an algorithm maps an ASM-IL$^+$ model into a set of MDGs (c.f., Figure 5.3 on page 68). An ASM-IL$^+$ model is given as a list of location-update pairs which describe the set of transition rules of the ASM model. Each pair describes the possible updates of one location. It will be transformed into an MDG. In order to do this, we have to treat the updates, the guards of the updates, and the else cases, which specify that the location is not changed if none of the guards are satisfied. In order to fully represent an ASM model, we also have to specify the domains of the locations and the initial state of the ASM. Moreover, a variable ordering that satisfies the well-formedness conditions should be suggested. This is described in this section.

Transforming updates of one location. Representing a location-update pair[7] $(loc_i, [(guard_{i1}, val_{i1}), \ldots])$ as an MDG is straightforward if we consider the equivalence (8.2). The next state variable loc'_i labels the root node of the graph. Each edge starting at the root is labelled with one of the specified values in the next state val_{ij} and leads to the subgraph G_{ij} that represents the corresponding guard of the update $guard_{ij}$. Figure 8.6 sketches a graph for a location-update pair.

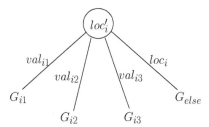

Figure 8.6: Location-update pair represented as MDG

Transforming the else-case. Figure 8.6 shows also that one edge of the MDG is labelled by the current value of the location loc_i and leads to a graph G_{else}. This branch specifies the else-case of the location-update pair: If none of the guards $guard_{ij}$ are satisfied, and therefore none of the updates can fire, the location should keep its value in the next state. For model checking, the semantics of the else-case has to be specified explicitly. Otherwise, the behaviour in this case is arbitrarily and the checker investigates every possible case.

For generating a branch that represents this else-case behaviour, we need to keep in mind that an edge cannot be labelled with a concrete variable. If loc_i is of abstract sort, then we may simply use the abstract state variable loc_i as an edge label. However, if loc_i is of concrete sort, then it has to be substituted by its current value val_{loc_i}. In this case, we have to generate a graph that comprises branches for all possible evaluations for the state variable loc_i. Thus, each branch represents the formula $loc'_i = val_i \wedge loc_i = val_i$, which obviously specifies that the location keeps its value. Figure 8.7 depicts an MDG that represents the else-case for a concrete location that ranges over three values, $\{val_1, val_2, val_3\}$. This (sub-)graph is disjoined with the MDG that represents the location-update pair (without else-case).

---

[7]For the readers' convenience we denote locational terms with loc and value terms as val.

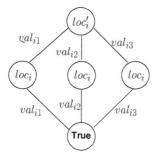

Figure 8.7: Representing the else-case for concrete locations

Transforming a guard. For generating the MDG G_{ij} that represents a guard, we use four basic functions (*and*, *or*, *negation*, and *equality*) as basic boolean operators that may appear in the guard (referred to as primary operators in the last section). Every other boolean operator can be mapped to these basic predicates.

- $and(opd_1, opd_2)$ is transformed into the **conjunct** of opd_1 and opd_2, assuming that both operands are given as MDGs.

- $or(opd_1, opd_2)$ is transformed into the **disjunct** of opd_1 and opd_2, if both operands are given as MDGs.

- $not(opd)$ is transformed into the **negation** of opd, if the operand is given as an MDG that does not contain node labels of abstract sort.

- $eq(lhs, rhs)$ assembles a new MDG. Its root is labelled by lhs, and it has a single edge labelled with rhs which leads to the leaf **True**.

Except for equations, the boolean operations take MDGs as operands. Thus, we recursively call the guard transformation function for the operands. The base of this recursion is the equality operator which operates on simple terms and constants and yields a simple graph.

Note that in most cases, the parameters of eq cannot be complex terms. Any parameter t_i of concrete sort will be simplified into a simple term $[\![t_i]\!]_\varsigma$ during the first step of the transformation that is introduced in the last section. These simple terms can be treated as labels. If one parameter is of abstract sort, the equation is mapped into a cross-term $isEq(lhs, rhs)$ and does not cause problems either (cross-terms can be used as labels as well).

The only terms that are not simplified by $[\![.]\!]_\varsigma$ are the primary operators for equality, conjunction and disjunction. If the operands of an equation are boolean expressions with non-simplified operators, we have to replace the equation $eq(lhs, rhs)$ with the expression $(lhs \wedge rhs) \vee (\neg lhs \wedge \neg rhs)$ which is logically equivalent[8].

Basic functions for conjoining and disjoining MDGs are given in the MDG-Package (see [Zho96]). Additionally, we implemented the algorithm for negation. This negation of an MDG is possible if no abstract variable appears as an lhs in any involved equations. As argued above, this is the case for guards $guard_{ij}$.

---

[8]Since guard expressions do not contain abstract variables on a LHS of an equation, negation is applicable.

Negation. Our algorithm for negation of an MDG M assumes that the root is not labelled by a variable of abstract sort (otherwise an error will be output). It works inductively on the structure of MDGs:

- if $M = \texttt{true}$ then $not(\texttt{true}) = \texttt{false}$

- if $M = \bigvee_{j \in J}(loc = val_j \wedge G_j)$

  where $dom(loc) = \{val_i \mid i \in I\}$, and the index set I, and the index subsets J and $\bar{J}$ of occurring and non-occurring indices are such that $J \subseteq I, \bar{J} \subseteq I, J \cap \bar{J} = \emptyset$, and $J \cup \bar{J} = I$ then the negation of M can be reduced to

  $$not(M) = \bigvee_{k \in \bar{J}}(loc = val_k \wedge \texttt{true}) \vee \bigvee_{j \in J}(loc = val_j \wedge not(G_j)).$$

  Figure 8.8 shows the sketch of the corresponding graphs for $I = \{val_1, val_2, val_3\}$.

Figure 8.8: Computing negation

Initial state, domains and variable ordering. The initial state of an ASM model is defined together with the declaration of functions and their domains. Every dynamic or external function may have a defined initial value. For our transformation we gather all initialisation information and put it into a formula that determines the set of initial states: $init\_state = \bigwedge_i(loc_i = init\_val_i)$. This conjunction need not be exhaustive since initialisation for some functions may be omitted. This formula is transformed into a simple MDG that consists of a single branch comprising all conjuncts.

The domains of an ASM model are defined as enumerations or by set comprehensions in the case of concrete sorts. Set comprehensions are unfolded by our transformation algorithm into enumerated sorts. When transforming the enumeration into an enumerated set for MDGs, we have to order the elements alphabetically. Otherwise, the internal substitution function will not work properly. For abstract sorts, we introduce a keyword that indicates a sort as being abstract.

Each location and each cross-term operator has an attached number in order to define a variable ordering. This ordering determines the shape of the corresponding MDG later on and heavily influences the graph size. We do not have a proper heuristic for a good ordering implemented yet[9]. Our implementation so far orders the locations and cross-term operators such that cross-term operators have a greater number than locations (in most cases the former depend on some of the latter and thus must have a greater ordering number). To support the user, we automatically generate a function which prints the list of locations and cross-term operators that are contributing to the current model. The order of this variable list can changed manually. Another function takes the newly ordered variable list as its input and adds a proper order-relation to the database that is used by the MDG algorithms.

Given this interface from ASM-IL$^+$ to the MDG-Package, and with respect to

---

[9]There is some ongoing work at the University of Montreal investigating heuristics for a good variable ordering for MDGs. The results are not published yet.

certain limitations for applying abstraction (as discussed in the last section), we are able to represent ASM models with abstract sorts by means of MDGs. This MDG representation of our model is thus treatable by MDG-based checking algorithms.

The work that is introduced in this chapter provides an interface from the ASM Workbench to the MDG-Package. An implementation of the algorithms is available. The code is given in standard ML in order to ease the interface to the ASM Workbench which is also implemented in ML. The program outputs a set of Prolog rules for generating MDGs that represent the ASM model. This includes apart from the initial state and the transition rules, the definition of sorts, variables and functions. These Prolog rules can easily be used as input for checking algorithms based on the MDG-Package that is also encoded in Prolog.

Our interface is tested only by using our own re-implementation of the reachability algorithm based on the MDG library of Prolog functions which is suggested in [CZS+97]. A complete model checking tool for MDGs is not available yet. However, we hope to get inspirations from the MDG-HDL model checker that is introduced in [Xu99, XCS+98] for our future development (see the discussion in Section 9.2).

Chapter 9

Conclusion

To summarise this thesis, we recall our initial intention: We aimed to provide tool support for analysing Abstract State Machines by means of model checking.

For the sake of a self-contained thesis, we recalled in the beginning the foundation of the formalism and the techniques that are used throughout this thesis: The syntax and semantics of *Abstract State Machines* (ASM) was given in Chapter 2. As a tool framework for editing, type-checking and simulating a typed version of ASM, called ASM-SL, we used the *ASM Workbench*. A brief description of this workbench completed this chapter. Chapter 3 gave an overview of model checking. In particular, we focused on *symbolic model checking* for the branching time logic CTL and the linear time logic LTL since these approaches appeared useful for the analysis of ASM. In Chapter 4, we described the data structure of *decision diagrams* that is the backbone of symbolic model checking.

Chapter 5, the centre of the thesis, illustrated how we think model checking can be useful for the development of complex systems. Basically, we did not aim to provide full verification support – model checking is much too restricted for such an ambitious enterprise – but support for *debugging* that is fully automatic and that nicely fits into a cyclic development process: We automatically transform the ASM model into the model checker language, specify the requirements to be checked, run the model checker, and simulate the counterexample that is output by the checking tool in case of an error. The simulation (or mere inspection) helps to get a better understanding of the behaviour of the ASM model and facilitates the correction of errors.

The second part of Chapter 5 described the interface between ASM and a checking tool. Facilitating ASM with model checking involves a language transformation: ASM have to be mapped into the input language of the model checker that is used. Since a variety of model checkers are available, of which each has its own strength for a particular application, it seems reasonable to provide more than one interface to one particular tool. Therefore, we intended to develop a general interface that can be used for many tools. Based on this design decision, we introduced an intermediate language, called ASM-IL, and split the transformation into two steps: In the first step, the transformation maps ASM-SL into ASM-IL. In a second step, ASM-IL is mapped to a particular model checker language. Since ASM-IL is a very simple formalism for describing transition systems, it can be easily transformed into many low-level tool languages. In Section 5.2, we introduced the technical details of the transformation from ASM into ASM-IL. We completed this section with a correctness proof based on the ASM semantics as given in Chapter 2.

Based on our general intermediate language, we provided interfaces to two model checking frameworks: We interfaced the ASM Workbench with the SMV tool ([McM93]) and with the MDG-Package ([CZS+97, CCL+97]).

Chapter 6 presented the interface to the SMV tool. We gave an overview of the tool and its input language as far as it is relevant for our approach. In Section 6.2, we described the transformation from ASM-IL into the SMV language. This interface is implemented in standard ML. It automatically transforms ASM-SL models into SMV code. In Chapter 7, we reported our experiences of using this interface for analysing two case studies: the FLASH cache coherence protocol ([KOH+94, Hei95]) and the Production Cell ([LL95]). While the former case study nicely showed the benefit of model checking, the latter one displayed its limitations.

The ASM model of the Production Cell that is investigated in this work is fairly simple because the system is modelled on a very abstract level: all domains of the model are finite. However, due to the fact that the Production Cell is an *embedded system* that strongly depends on the environment's behaviour, the model becomes much more complex if it is analysed with respect to its environment. Since model checking is an exhaustive search in the state space, the tool checks every possible behaviour. This includes also the possible behaviour of the sensor variables that define the environment. We needed to restrict the behaviour of the sensor variables in order to treat the model by the SMV tool, i.e., we added a model of the *assumptions* made on the environment. Choosing an environment model that is too restrictive, however, entails the danger of missing possible errors. This experience, it is discussed in detail in Section 7.2, nicely motivated support towards two directions: Firstly, modelling assumptions on the environment can be nicely done in a temporal logic. We need to provide a model checker that allows the user to add temporal logic formulas to the system model. As described in Section 3.4, this can be done in the framework of LTL model checking which is based on *tableau construction*. Secondly, we need a means to reduce complexity. *Abstraction* is one technique for extending the limits of model checking towards the treatment of more complex models.

The last chapter, Chapter 8, treated both issues. It described the interface from ASM to *multiway decision graphs* (MDGs). MDGs, as described in Section 4.2, support abstract sorts and functions. Moreover, model checking algorithms for checking a subset of first order ACTL* (the universal fragment of CTL*) are already provided in the literature (see [Xu99, XCS+98]). These algorithms involve a tableau construction similar to the approach for LTL checking. For these reasons, we developed an interface to the MDG framework. In Section 8.3, we explained how the transformation allows the user to automatically generate abstract models once the abstract sorts are distinguished. This support takes the burden of manually re-modelling the system model from the user if the model is too complex and must be abstracted. A transformation from ASM into MDGs was described in Section 8.4. Since MDGs provide a means for modelling abstract sorts, we had to extend the first step of the transformation with a feature for treating abstract functions that are left *uninterpreted*.

Our second interface from ASM to MDGs is implemented. However, an *MDG model checker* for our needs is not yet developed. This is discussed as future work in Section 9.2.

9.1 Related Work

We discuss the work that is related to ours from two different viewpoints: On one hand, we discuss other approaches for model checking Abstract State Machines. On the other hand, we reflect on other techniques for supporting abstraction. This section is divided into two corresponding subsections.

9.1.1 Model Checking Abstract State Machines

Extending tool environments for high-level specification languages with an interface
to a model checker is an upcoming topic. One can find approaches that are quite
similar to ours but work on a different language: [Day93] suggests a transformation
from Statecharts into SMV; in [LR98], controller specifications, modelled in the
language CSL, are transformed and model checked by SVE ([F+96]); the work that
is introduced in [MMP98] equips the multi-language environment SYNCHRONIE with
an interface to the VIS model checker, and so on.

Closer to our approach from the language point of view is the work by Spielmann. In [Spi99], he investigates automatic verification of ASM with unbounded
input. This is done by representing an ASM model "symbolically" by means of
a logic for computation graphs (called CGL*). The resulting formula is combined
with a CTL*-like formula which specifies properties and checked by means of de-
ciding its finite validity. This approach addresses the problem of checking systems
with infinitely many inputs but it is only applicable to ASM with 0-ary dynamic
functions and input that is restricted to relations[1]. Spielmann proves that the
decision procedure is PSPACE-complete and optimal. In his framework, the verification of generalised nullary programs which have functions in their input (instead of
relations only) becomes undecidable.

From the theoretical point of view, this work provides very interesting results
that could help to support the analysis of embedded systems that involve unbounded
non-determinism. We hope for further development of this approach that extends
its limitations. The conclusion of [Spi99] already suggests the use of fragments of
the logic CGL* in order to provide decidability of the algorithms that can check
validity for a broader range of ASM models.

As a more practical approach, Plonka develops in his diploma thesis ([Plo00])
an interface from ASM to the model checker SVE ([F+96]). The transformation
schema that is used to map Abstract State Machines into the interchange format,
called RSM, is quite similar to the transformation that is introduced in this work.
This strengthens our claim that our interface language ASM-IL provides generality
and can be used for interfacing many tools. As far as we know, the transformation
algorithm is not yet implemented. The results of [Plo00], however, could be used to
build a third interface from ASM-IL to the model checker SVE. This model checker
implements the LTL model checking approach that is described in Section 3.4. It
allows the user to introduce temporal logic assumptions into the system model to
be checked. However, the tool does not provide support for abstraction.

9.1.2 Abstraction Techniques

Abstraction is sometimes referred to as the key methodology for combining deductive and algorithmic techniques. The basic idea behind abstraction is to reduce the
size of a possibly infinite model. When successful, abstraction provides an abstract
version of the concrete model which eases the verification task. Due to the limitation of automatic approaches for supporting verification, e.g., model checkers or
validity checkers, a successful abstraction technique can be crucial to the success of
algorithmic verification.

Often in the literature, the process of abstracting is described in terms of separating the control flow of a model from the data part and neglecting the latter. This
viewpoint is introduced in the context of hardware verification. The basic problem
is that this separation is not obvious whenever data influences the control flow.

---

[1]In contrast to ASM as described in Chapter 2, Spielmann allows relation as well as function
symbols in his signature.

As a suitable solution for a certain application domain, Hojati, Dill and Brayton introduce in [HDB97] an approach for checking integer-based circuit models. In certain models, integer variables can be replaced by enumeratable variables. Thus, the model turns into a finite instance and can be treated by a BDD-based model checker. This technique, however, is not applicable for abstracting arbitrary data: Instead of building a finite instance, a BDD-based reachability procedure checks the first n steps of the circuit behaviour against a given invariant. A full state exploration, however, is not applicable.

Clarke, Grumberg and Long ([CGL92, Lon93]) introduce a technique for abstracting data in a complex circuit design. The width of the data path is reduced and thus the abstract model is much smaller. BDD-based model checking becomes applicable. To apply this technique, an *abstraction function* α has to be defined by the designer. Moreover, the correctness of the abstract model with respect to the original model is not always obvious. It has to be proven mechanically. Both tasks, providing an α and proving correctness are quite complex and cannot be supported by fully algorithmic tools.

In the same context, Saïdi, Graf, and Shankar ([GS97, SS99]) propose a framework for supporting the computation of the abstract model and the abstract property to be checked, once the abstraction function α is defined, and for proving correctness of the abstract model by means of the theorem prover PVS ([ORS92]). The theoretical foundation is introduced in [LGS+95] as a *property preserving* abstraction technique.

Hungar, Grumberg and Damm in [HGD95] introduce the notion of "truly symbolic" representation in contrast to symbolic representation introduced in [McM93]. This approach separates the control part of the system from its data part and refines the resulting model into an intermediate description that contains sufficient information about the property to be checked. The enrichment of the control flow model introduces additional proof obligations, so called *verification conditions*. These verification conditions can be generated automatically. The requirements to be checked are modelled in first-order ACTL, the universal fragment of CTL[2]. Data values and operations can be modelled in this logic by means of first-order predicates. If a given requirement does not contain predicates on data, then BDD model checking is applicable. Otherwise, all first-order predicates are substituted by *true*. If the model checking procedure reports a failure, this failure will be part of the control flow and occurs in the concrete model as well. If no failure can be found, correctness is not guaranteed in the original model since failure in the abstracted data part cannot be detected. Therefore, the generated verification conditions have to be proven. This is done by means of a theorem prover. In contrast to other approaches, no abstraction function has to be defined; only the control variables have to be identified. However, the verification conditions have to be proven manually. Moreover, this approach is restricted to circuits with terminating data loops and a clear separation between data and control part.

Bharadwaj and Heitmeyer suggest in [BH99] algorithmic support for deriving an abstract model from the requirements to be checked. They analyse the abstract model (with respect to the abstracted requirements) by means of model checkers and translate, if an error is detected, the counterexample back into the concrete model. Their approach is derived from techniques that are used in practice when a model is manually abstracted in an ad hoc way. (The common practice for abstracting manually is also described in [WVF97].) They describe two methods for deriving abstraction. The first method simply eliminates irrelevant variables, i.e., variable that have no influence for the requirements to be checked. For example, in the

---

[2]That is, ACTL is a restricted sub-logic of CTL that does not comprise existential quantification over paths denoted by the E operator.

FLASH case study that is described in Section 7.1, the variable *data* has no influence on the protocol and can be eliminated. The second method allows a *monitored* variable[3] v to be removed if it does not occur in the requirement q but influences a variable v' that occurs in q. In the abstract model, v' is treated as a monitored variable. In order to reduce the domain of monitored variable v', a mapping is introduced that partitions the original domain of v into equivalence classes. The resulting abstract model is thus smaller.

The second method is partly comparable with the techniques we use when transforming ASM into MDG (see Section 8.4.1). However, it is more restricted since it is applicable only if v' is the only variable that depends on that eliminated. Furthermore, a mapping has to be defined that yields appropriate equivalence classes. A similar partitioning is given automatically by our transformation when introducing new cross-terms.

Bharadwaj and Heitmeyer show that their abstraction techniques are *sound* and *complete*. In contrast to that, our approach is sound but not complete. Completeness assures that every counterexample that can be found in the abstract model is also a counterexample in the concrete model. That is, wrong counterexamples (as discussed in Section 8.3) do not occur.

In contrast to problem reduction by means of computing an abstract model, other approaches suggest tools that are based on different structures for representing a system. These approaches are more closely related to the MDG approach that is used in this thesis. Uninterpreted functions and predicates are introduced into various frameworks.

Burch and Dill ([BD94]) investigated the pipelined DLX architecture by means of modelling data values and operations with functions that are left uninterpreted during the checking process. Similar to the MDG approach, this technique allows systems that can be arbitrarily large in their data parts to be treated. In contrast to the MDG approach, however, they support verification by means of validity checking procedures. As a consequence, no temporal properties, such as liveness, can be checked. This is possible with the model checking algorithm for MDGs introduced in [Xu99, XCS+98].

In [CN94], Cyrluk and Narendran introduce a new logic, called ground temporal logic (GTL). This logic also allows uninterpreted function symbols. They show that a decidable fragment of GTL can be treated by an automatic validity checker (based on PVS). The thesis of Xu ([Xu99]) that introduces the logic $\mathcal{L}_{MDG}$ and the corresponding model checking algorithm based on MDGs shows that the decidable fragment of GTL is a subset of $\mathcal{L}_{MDG}$. Moreover, the MDG model checking approach goes beyond validity checking.

9.2 Directions for Future Work

As one future direction, we intend to complete the implementation of the interface from the ASM Workbench to the SMV tool. Two tasks are left to do:

1. The re-transformation from the counterexample that is output by the SMV tool into the simulator of the workbench has to be implemented. In principle, this interface is easily realised as discussed in Section 6.

2. Modelling of the requirements that are to be checked should be supported. For the use of SMV, the requirements must be specified in the temporal logic CTL. The user interface of the ASM Workbench should be extended to allow support for the user to generate these CTL specifications. For many safety

---

[3] These monitored variables are comparable with external functions in ASM.

and liveness properties, a schema for the corresponding CTL formula might be automatically suggested by the tool.

As another future direction, we plan the further development of a model checker based on MDGs that is tailored for our needs. For this development, we can use results that are given in the literature. [Xu99, XCS$^+$98] introduces a model checker based on MDGs. In this work, Xu, Cerny, Song and Corella invented an extension for CTL*, called *Abstract_CTL*. This logic allows the use of abstract sorts and first-order quantification. In her thesis ([Xu99]), Xu introduces a logical language called $\mathcal{L}_{MDG}$ which is a subset of Abstract_CTL*. This language is restricted to universal path quantification and a limited nesting of temporal operators. This restriction allows the development of model checking algorithms.

Similar to the approach described in Section 3.4, model checking of $\mathcal{L}_{MDG}$ is based on a tableau construction of the formula that is to be checked. In contrast to the LTL model checking approach in Section 3.4, the "tableaus" that are constructed in [Xu99] are not given as simple transition systems but as circuit models specified in MDG-HDL. These circuits derived from formulas are conjoined with the circuit specification to be checked which is also given in MDG-HDL. The composed circuit is analysed by the MDG-HDL model checker.

In our case, when model checking ASM, we need to develop a tableau construction that yields a tableau as a simple transition system rather than a piece of hardware description. This development of a model checker based on MDGs and tableau construction for simple transition systems is planned as future work in cooperation with Concordia University in Montreal.

Appendix A

Appendix

A.1   Fixpoint Characterisation

Lemma A.1. $f(Z) := \varphi \wedge \mathbf{EX}\, Z$ is monotonic, i.e., for sets of states S_i and S_j with $S_i \subseteq S_j$ it follows that $f(S_i) \subseteq f(S_j)$.

Proof: Consider $s \in f(S_i)$, using the definition of f we get $s \models \varphi$ and there is some other state s' with $(s, s') \in R$ that is also an element of S_i, so $s' \in S_i$. Since $S_i \subseteq S_j$ we have $s' \in S_j$ and so s is in the set characterised by $\varphi \wedge \mathbf{EX}\, S_j$, that is $s \in f(S_j)$. $\blacksquare$

Lemma A.2. $\mathbf{EG}\, \varphi$ is a fixpoint for the predicate transformer $f(Z) := \varphi \wedge \mathbf{EX}\, Z$.

Proof: Assume $s_0 \models \mathbf{EG}\, \varphi$. By definition of $\models$ and the $\mathbf{EG}$ operator there must be a path $s_0, s_1, s_2, \ldots$ in structure M such that for all $k \geq 0$, $s_k \models \varphi$. It follows that $s_0 \models \varphi$ as well as $s_1 \models \mathbf{EG}\, \varphi$. This implies $s_0 \models \varphi$ and $s_0 \models \mathbf{EX}\, \mathbf{EG}\, \varphi$, thus $\mathbf{EG}\, \varphi \subseteq \varphi \wedge \mathbf{EX}\, \mathbf{EG}\, \varphi$. On the other hand, with $s_0 \models \varphi \wedge \mathbf{EX}\, \mathbf{EG}\, \varphi$ we get $s_0 \models \mathbf{EG}\, \varphi$, and so $\mathbf{EG}\, \varphi \supseteq \varphi \wedge \mathbf{EX}\, \mathbf{EG}\, \varphi$. We conclude that $\mathbf{EG}\, \varphi = \varphi \wedge \mathbf{EX}\, \mathbf{EG}\, \varphi$. $\blacksquare$

Proving that $\mathbf{EG}\, \varphi$ is the greatest fixpoint we need an additional lemma for the predicate transformer $f(Z) = \varphi \wedge \mathbf{EX}\, Z$.

Lemma A.3. Let $f(Z) = \varphi \wedge \mathbf{EX}\, Z$ and $f^{\infty}(S)$ be the limit of the sequence $S \supseteq f(S) \supseteq f^2(S) \supseteq \ldots$ (i.e., the value after infinitely many iterations of applying f). Then for every $s \in S$, $s \in f^{\infty}(S)$ implies $s \models \varphi$, and there is a state $s' \in S$ with $(s, s') \in R$ and $s' \in f^{\infty}(S)$.

Proof: Because of Lemma A.1 we know that f has a fixpoint. This is given by the limit f^{∞}, so that $f^{\infty}(S) = f(f^{\infty}(S))$. By definition of f we get that $s \models \varphi$, and there is a state $s' \in S$ with $(s, s') \in R$ and $s' \in f^{\infty}(S)$. $\blacksquare$

Lemma A.4. $\mathbf{EG}\, \varphi$ is the greatest fixpoint for predicate transformer $f(Z) := \varphi \wedge \mathbf{EX}\, Z$, i.e., $\mathbf{EG}\, \varphi = fp_{max}(Z, \varphi \wedge \mathbf{EX}\, Z)$.

Proof: Since we are dealing with a complete lattice and f if monotonic (following Lemma A.1), f is $\cap$-continuous as well. It suffices to prove $\mathbf{EG}\, \varphi = \bigcap_i f^i(S)$ where S is the supremum of our lattice $\mathcal{P}(\mathcal{S})$.

We first show that $\mathbf{EG}\, \varphi \subseteq \bigcap_i f^i(S)$. By induction on i we establish $\mathbf{EG}\, \varphi \subseteq f^i(S)$ for all i. Obviously, $\mathbf{EG}\, \varphi \subseteq S$. For arbitrary n, let $\mathbf{EG}\, \varphi \subseteq f^n(S)$. Since f is monotonic it follows that $f(\mathbf{EG}\, \varphi) \subseteq f^{n+1}(S)$. By Lemma A.2, $f(\mathbf{EG}\, \varphi) = \mathbf{EG}\, \varphi$, and hence $\mathbf{EG}\, \varphi \subseteq f^{n+1}(S)$.

To show $\mathbf{EG}\, \varphi \supseteq \bigcap_i f^i(S)$ we assume state $s \in \bigcap_i f^i(S)$. This state is included in all $f^i(S)$. Hence it is also included in the fixpoint $f^{\infty}(S)$. With Lemma A.3, s is the start of an infinite sequence of states in M in which each state is related to the previous state by relation R. Furthermore, each of these states satisfies φ. Thus, $s \models \mathbf{EG}\, \varphi$. $\blacksquare$

The proof for the fixpoint characterisation of $\mathbf{A}\, (\varphi\ \mathbf{U}\ \psi)$ follows the same steps. We first show the monotonicity of the transformer, then proving that the $\mathbf{A}\, (\varphi\ \mathbf{U}\ \psi)$ is a fixpoint and for the last step it remains to show that it is the least fixpoint.

Lemma A.5. $f(Z) := \psi \vee (\varphi \wedge \mathbf{AX}\, Z)$ is monotonic, i.e., for sets of states S_i and S_j with $S_i \subseteq S_j$, $f(S_i) \subseteq f(S_j)$ follows.

Proof: Assume $S_i \subseteq S_j$. If $s \in f(S_i)$, then using the definition of f, $s \models \psi$ or $s \models \varphi$ and for all state s' with $(s, s') \in R$, $s' \models \mathbf{A}\, (\varphi\ \mathbf{U}\ \psi)$, i.e., also $s' \in S_i$ and because of the subset relation $s' \in S_j$. Thus, $s \models \psi \vee (\varphi \wedge \mathbf{AX}\, S_j)$ as well, that is $s \in f(S_j)$. $\blacksquare$

Lemma A.6. $\mathbf{A}(\varphi \ \mathbf{U} \ \psi)$ *is a fixpoint for the predicate transformer* $f(Z) := \psi \vee (\varphi \wedge \mathbf{AX} \ Z)$.

Proof: We prove the lemma in two steps: $\mathbf{A}(\varphi \ \mathbf{U} \ \psi) \subseteq \psi \vee (\varphi \wedge \mathbf{AX}(\mathbf{A}(\varphi \ \mathbf{U} \ \psi)))$ and $\mathbf{A}(\varphi \ \mathbf{U} \ \psi) \supseteq \psi \vee (\varphi \wedge \mathbf{AX}(\mathbf{A}(\varphi \ \mathbf{U} \ \psi)))$.

Assume $s_0 \models \mathbf{A}(\varphi \ \mathbf{U} \ \psi)$. According to the semantics given in definition 3.16 it follows that for all path $s_0, s_1, s_2, \ldots$ there is a $k \geq 0$ with $s_k \models \psi$ and for all j with $0 \leq j < k$, $s_j \models \varphi$. Thus, $s_0 \models \psi$ or $s_0 \models \varphi$ and for all path $s_1, s_2, \ldots$ (that is, all paths starting at a successor of s_0) there is a $k \geq 1$ with $s_k \models \psi$ and for all j with $1 \leq j < k$, $s_j \models \varphi$. In terms of CTL this means $s_0 \models \psi \vee (\varphi \wedge \mathbf{AX}(\mathbf{A}(\varphi \mathbf{U} \ \psi)))$. We get $\mathbf{A}(\varphi \ \mathbf{U} \ \psi) \subseteq f(\mathbf{A}(\varphi \ \mathbf{U} \ \psi))$.

For proving $\mathbf{A}(\varphi \ \mathbf{U} \ \psi) \supseteq \psi \vee (\varphi \wedge \mathbf{AX}(\mathbf{A}(\varphi \ \mathbf{U} \ \psi)))$ we assume that $s_0 \models \psi \vee (\varphi \wedge \mathbf{AX}(\mathbf{A}(\varphi \ \mathbf{U} \ \psi)))$. This means that $s_0 \models \psi$ or $s_0 \models \varphi$ and for s_1 with $(s_0, s_1) \in R$ (i.e., all next states) $s_1 \models \mathbf{A}(\varphi \ \mathbf{U} \ \psi)$. This is equivalent with the statement that $s_0 \models \psi$ or $s_0 \models \varphi$ and for all paths $s_0, s_1, s_2, \ldots$ there is a $k \geq 1$, $s_k \models \psi$ and for all $1 \leq j < k$, $s_j \models \varphi$. Both cases can be conjoined: for all path $s_0, s_1, s_2, \ldots$ there is a $k \geq 0$, $s_k \models \psi$ and for all $0 \leq j < k$, $s_j \models \varphi$, which is equivalent to $s_0 \models \mathbf{A}(\varphi \ \mathbf{U} \ \psi)$. ∎

Lemma A.7. $\mathbf{A}(\varphi \ \mathbf{U} \ \psi)$ *is the least fixpoint for predicate transformer* $f(Z) := \psi \vee (\varphi \wedge \mathbf{AX} \ Z)$, *i.e.*, $\mathbf{A}(\varphi \ \mathbf{U} \ \psi) = fp_{min}(Z, \psi \vee (\varphi \wedge \mathbf{AX} \ Z)$.

Proof: Since we are dealing with a complete lattice and f is monotonic (following Lemma A.5), f is $\cup$-continuous as well. Following equation (3.4) it suffices to prove $\mathbf{A}(\varphi \ \mathbf{U} \ \psi) = \bigcup_i f^i(\emptyset)$ where $\emptyset$ is the infimum of our lattice $\mathcal{P}(\mathcal{S})$.

It is easy to see that $f^i(\emptyset) \subseteq \mathbf{A}(\varphi \ \mathbf{U} \ \psi)$ for all i when looking at the results of iteratively applied f:

$$
\begin{aligned}
f^0(\emptyset) &= \emptyset \\
f^1(\emptyset) &= \psi \vee (\varphi \wedge \mathbf{AX} \ \emptyset) &= \psi \\
f^2(\emptyset) &= \psi \vee (\varphi \wedge \mathbf{AX}(f^1(\emptyset))) &= \psi \vee (\varphi \wedge \mathbf{AX} \ \psi) \\
f^3(\emptyset) &= \psi \vee (\varphi \wedge \mathbf{AX}(f^2(\emptyset))) &= \psi \vee (\varphi \wedge \mathbf{AX}(\psi \vee (\varphi \wedge \mathbf{AX} \ \psi))) \\
f^4(\emptyset) &= \psi \vee (\varphi \wedge \mathbf{AX}(f^3(\emptyset))) &= \psi \vee (\varphi \wedge \mathbf{AX}(\psi \vee (\varphi \wedge \mathbf{AX}(\psi \vee (\varphi \wedge \mathbf{AX} \ \psi))))) \\
f^5(\emptyset) &= \ldots
\end{aligned}
$$

We can see that the iterated application enlarges the length of the maximal prefix of the paths satisfying φ until ψ is satisfied, which is given by the upper bound for parameter k in the semantic definition. For $\mathbf{A}(\varphi \ \mathbf{U} \ \psi)$, however, k is unbound. Thus, $f^i(\emptyset) \subseteq \mathbf{A}(\varphi \ \mathbf{U} \ \psi)$ for all i, and so is $\bigcup_i f^i(\emptyset) \subseteq \mathbf{A}(\varphi \ \mathbf{U} \ \psi)$.

$\mathbf{A}(\varphi \ \mathbf{U} \ \psi) \subseteq \bigcup_i f^i(\emptyset)$ can be proved by induction over parameter k in the semantical definition. The induction basis $s_0 \models \mathbf{A}(\varphi \ \mathbf{U} \ \psi)$ for a $k = 0$ implies $s_0 \models \psi$ so that $s_0 \in f^1(\emptyset)$. As the induction hypothesis we assume that $s_0 \models \mathbf{A}(\varphi \ \mathbf{U} \ \psi)$ and $0 \leq k \leq (n-1)$ implies that $s_0 \in f^n(\emptyset)$. For the inductive step let k be bound to n: $s_0 \models \mathbf{A}(\varphi \ \mathbf{U} \ \psi)$ and there is a k with $0 \leq k \leq n$ such that $s_k \models \psi$ and for all $0 \leq j < k$, $s_j \models \varphi$. Considering now $s_0 \models \varphi$ and all paths $s_1, s_2, \ldots$ with $(s_0, s_1) \in R$ (i.e., all paths starting at a successor of s_0). These are the suffixes of length $(n-1)$ satisfying φ until ψ is satisfied. With the induction hypothesis it follows that $s_1 \in f^n(\emptyset)$. Since $(s_0, s_1) \in R$ and $s_0 \models \mathbf{A}(\varphi \ \mathbf{U} \ \psi)$ it follows that $s_0 \models \psi \vee (\varphi \wedge \mathbf{AX}(f^n(\emptyset)))$. Thus, $s_0 \in f(f^n(\emptyset)) = f^{n+1}(\emptyset)$. We conclude that $\mathbf{A}(\varphi \ \mathbf{U} \ \psi) \subseteq f^i(\emptyset)$ for all i, and since f is $\cup$-continuous $\mathbf{A}(\varphi \ \mathbf{U} \ \psi) \subseteq \bigcup_i f^i(\emptyset)$. ∎

A.2 FLASH Cache Coherence Protocol

A.2.1 The ASM-SL Code

```
(*-----------------------------------------------------------------*)

freetype  TYPE    == { noMess, get, getx, inv,  wb, rpl, fwdack, swb,
                       invack,  nack, nackc, fwdget, fwdgetx, put,
                       putx, nackc2, put_swb, putx_fwdack }

freetype  CCTYPE  == { ccget, ccgetx, ccrpl, ccwb }

freetype  AGENT   == { agent :INT, none }
freetype  LINE    == { lines :INT }

typealias MESSAGE == AGENT * TYPE * AGENT * LINE

freetype PHASE    == { ready, wait, invalidPhase }
freetype STATE    == { exclusive, shared, invalid }
freetype LENGTH   == { n :INT }

(*------------------------------------- *)

static function max_agent == 2
static function max_line  == 2
static function maxQ  == 2

static function Agent == {agent(i)| i in {1..max_agent}} union {none}
static function Agent_without_none == {agent(i)| i in {1..max_agent}}
static function Line  == { lines(i) | i in {1..max_line} }
static function QLength  == { n(i) | i in {1..maxQ} }

(*-----------------------------------------*)

static function Type  == { noMess, get, getx, inv,  wb, rpl, fwdack,
                           swb, invack,  nack, nackc, fwdget,
                           fwdgetx, put, putx, nackc2, put_swb,
                           putx_fwdack }

static function CCType  == { ccget, ccgetx, ccrpl, ccwb }
static function ReqType == { noMess, get, getx, rpl, wb }


static function Home ==
  MAP_TO_FUN { lines(1) -> agent(1), lines(4) -> agent(1),
               lines(2) -> agent(1), lines(5) -> agent(2),
               lines(3) -> agent(3), lines(6) -> agent(3) }

(*-----------------------------------------*)

dynamic function MessInTr : LENGTH * AGENT -> TYPE
  with MessInTr(i,a) in Type
  initially MAP_TO_FUN {(i,a) -> noMess | (i,a) in
           Union({ {(i,a) | i in QLength } | a in Agent\{none}}) }
```

```
dynamic function SenderInTr : LENGTH * AGENT -> AGENT
   with SenderInTr(i,a) in Agent \ { none }
   initially MAP_TO_FUN { (i,a) -> agent(1) | (i,a) in
        Union({ { (i,a) | i in QLength } | a in Agent \ { none }}) }

dynamic function SourceInTr : LENGTH * AGENT -> AGENT
  with SourceInTr(i,a) in Agent \ { none }
  initially MAP_TO_FUN { (i,a) -> agent(1) | (i,a) in
        Union({ { (i,a) | i in QLength } | a in Agent \ { none }}) }

dynamic function LineInTr : LENGTH * AGENT -> LINE
  with LineInTr(i,a) in Line
  initially MAP_TO_FUN { (i,a) -> lines(1) | (i,a) in
        Union({ { (i,a) | i in QLength } | a in Agent \ { none }}) }

(*----------------------------------------*)

dynamic function SenderInTrR : AGENT -> AGENT
  with SenderInTrR(a) in Agent \ { none }
  initially MAP_TO_FUN { a -> agent(2) | a in Agent \ { none } }

dynamic function SourceInTrR : AGENT -> AGENT
  with SourceInTrR(a) in Agent \ { none }
  initially MAP_TO_FUN { a -> agent(1) | a in Agent \ { none } }

dynamic function MessInTrR : AGENT -> TYPE
  with MessInTrR(a) in ReqType
  initially MAP_TO_FUN { a -> noMess | a in Agent \ { none } }

dynamic function LineInTrR : AGENT -> LINE
  with LineInTrR(a) in Line
  initially MAP_TO_FUN { a -> lines(1) | a in Agent \ { none } }


(*----------------------------------------*)

dynamic function InSender : AGENT -> AGENT
  with InSender(a) in Agent \ { none }
  initially MAP_TO_FUN { a -> agent(2) | a in Agent \ { none } }

dynamic function InSource : AGENT -> AGENT
  with InSource(a) in Agent \ { none }
  initially MAP_TO_FUN { a -> agent(2) | a in Agent \ { none } }

dynamic function InMess : AGENT -> TYPE
  with InMess(a) in Type
  initially MAP_TO_FUN { a -> noMess | a in Agent \ { none } }

dynamic function InLine : AGENT -> LINE
  with InLine(a) in Line
  initially MAP_TO_FUN { a -> lines(1) | a in Agent \ { none } }
(*----------------------------------------*)
```

```
dynamic function CurPhase :AGENT * LINE -> PHASE
   initially MAP_TO_FUN
   { (a,l) -> ready | (a,l) in Union ({ { (a,l) |
                     a in Agent \ { none }} | l in Line} )}

dynamic function CCState :AGENT * LINE -> STATE
   initially MAP_TO_FUN { (a,l) -> invalid | (a,l) in
       Union ({ { (a,l) | a in Agent \ { none }} | l in Line} )}

dynamic relation Pending  :LINE
   initially SET_TO_REL {}

dynamic function Owner    :LINE -> AGENT
   with Owner (l) in Agent
   initially MAP_TO_FUN { l -> none | l in Line }

dynamic relation Sharer   :LINE * AGENT
   initially SET_TO_REL {}


(*-----------------------------------------*)
(*-- synchronization ----------------------*)

freetype MODE   == { behave, sync }

dynamic function toggle : MODE
   initially behave

(*-----------------------------------------*)

external function Self    :AGENT
  with Self in Agent \ { none }

external function produce :AGENT -> (CCTYPE * LINE)
    with produce(a) in
    Union({{(cctype_, line_) | cctype_ in CCType } | line_ in Line})

(*-----------------------------------------------------------------*)


transition AppendToTransit (agent_, (sender_,mess_,source_,line_)) ==
   if MessInTr(n(1),agent_)=noMess
   then  SenderInTr(n(1),agent_) := sender_
         MessInTr(n(1),agent_)   := mess_
         SourceInTr(n(1),agent_):= source_
         LineInTr(n(1),agent_)   := line_
   else do forall i in { 2..maxQ }  (* the first empty Queue-entry *)
         if MessInTr(n(i-1),agent_)!=noMess
            and MessInTr(n(i),agent_)=noMess
         then SenderInTr(n(i),agent_) := sender_
              MessInTr(n(i),agent_)   := mess_
              SourceInTr(n(i),agent_):= source_
              LineInTr(n(i),agent_)   := line_
         endif
```

```
        enddo
    endif


transition AppendRequestToTransit (agent_,
                                    (sender_,mess_,source_,line_)) ==
  if MessInTrR (agent_)=noMess
  then SenderInTrR(agent_) := sender_
       MessInTrR (agent_)  := mess_
       SourceInTrR (agent_):= source_
       LineInTrR (agent_)  := line_
  endif


(*----------------------------------------------------------------*)

transition R_Requests ==
  if MessInTrR(agent(1)) = noMess
  then
    case produce(Self) of
      (ccget, 1) :
        if CurPhase(Self, 1)=ready
        then AppendRequestToTransit (Home(1),(Self,get,Self,1))
        endif;
      (ccgetx,1) :
        if CurPhase(Self, 1)=ready
        then AppendRequestToTransit (Home(1), (Self,getx,Self,1))
        endif;
      (ccrpl,1) :
         if CurPhase(Self, 1)=ready and CCState(Self,1)=shared
         then AppendRequestToTransit (Home(1),(Self,rpl,Self,1))
         endif;
       (ccwb,1) :
         if CurPhase(Self, 1)=ready and CCState(Self,1)=exclusive
         then AppendRequestToTransit (Home(1),(Self,wb,Self,1))
         endif;
      _ : skip
    endcase
  endif

(*----------------------------------------------------------------*)

transition R5 ==
  if (InMess(Self)=get and Home(InLine(Self)) = Self)
  then
     if Pending (InLine(Self))
      then if  MessInTr(n(maxQ),InSource(Self))=noMess
           then AppendToTransit (InSource(Self),
                        (Self,nack,InSource(Self),InLine(Self)))
                InMess(Self):=noMess
           endif
      else if Owner (InLine(Self)) != none
            then if MessInTr(n(maxQ),Owner(InLine(Self)))=noMess
                 then AppendToTransit (Owner (InLine(Self)),
```

```
                        (Self,fwdget,InSource(Self),InLine(Self)))
                Pending (InLine(Self)) := true
                InMess(Self):=noMess
            endif
      else if MessInTr(n(maxQ),InSource(Self))=noMess
          then AppendToTransit (InSource(Self),
                (Self,put,InSource(Self),InLine(Self)))
                InMess(Self):=noMess
                Sharer (InLine(Self), InSource(Self)) := true
            endif
      endif
    endif
  endif


transition R6 ==
  if InMess(Self) = fwdget
  then if CCState (Self,InLine(Self)) = exclusive
      then if (Home(InLine(Self))=InSource(Self))
          then if MessInTr(n(maxQ),Home(InLine(Self))) = noMess
              then AppendToTransit ( Home(InLine(Self)),
                    (Self,put_swb,InSource(Self),InLine(Self)))
                    CCState(Self,InLine(Self)):=shared
                    InMess(Self):=noMess
              endif
          else if MessInTr(n(maxQ),InSource(Self)) = noMess
                and MessInTr(n(maxQ),Home(InLine(Self))) = noMess
              then AppendToTransit (InSource(Self),
                      (Self,put,InSource(Self),InLine(Self)))
                    AppendToTransit ( Home(InLine(Self)),
                        (Self,swb,InSource(Self),InLine(Self)))
                    CCState(Self,InLine(Self)):=shared
                    InMess(Self):=noMess
              endif
          endif
      else if (Home(InLine(Self))=InSource(Self))
          then if MessInTr(n(maxQ),Home(InLine(Self))) = noMess
              then AppendToTransit (Home(InLine(Self)),
                    (Self,nackc2,InSource(Self),InLine(Self)))
                    InMess(Self):=noMess
              endif
          else if MessInTr(n(maxQ),InSource(Self)) = noMess
                and MessInTr(n(maxQ),Home(InLine(Self))) = noMess
              then AppendToTransit (InSource(Self),
                      (Self,nack,InSource(Self),InLine(Self)))
                    AppendToTransit (Home(InLine(Self)),
                        (Self,nackc,InSource(Self),InLine(Self)))
                    InMess(Self):=noMess
              endif
          endif
      endif
    endif
  endif

transition R7 ==
```

```
 if InMess(Self) = put
 then if CurPhase(Self,InLine(Self)) != invalidPhase
      then CCState(Self,InLine(Self)) := shared
      endif
      CurPhase(Self,InLine(Self)) := ready
      InMess(Self) := noMess
  endif

transition R8 ==
  if (InMess(Self) = swb and Home(InLine(Self)) = Self)
  then
      Sharer(InLine(Self), InSource(Self)) := true
      if Owner(InLine(Self)) != none
      then Sharer(InLine(Self), Owner(InLine(Self))) := true
      endif
      Owner(InLine(Self)) := none
      Pending(InLine(Self)) := false
      InMess(Self) := noMess
  endif
```

(*--*)

```
transition R7_R8 ==
 if InMess(Self) = put_swb
 then if CurPhase(Self,InLine(Self)) != invalidPhase
         then CCState(Self,InLine(Self)) := shared
      endif
      CurPhase(Self,InLine(Self)) := ready
      Sharer(InLine(Self), InSource(Self)) := true
      if Owner(InLine(Self)) != none
         then Sharer(InLine(Self), Owner(InLine(Self))) := true
      endif
      Owner(InLine(Self)) := none
      Pending(InLine(Self)) := false
      InMess(Self) := noMess
  endif
```

(*--*)

```
transition R9 ==
  if InMess(Self) = nack
  then
      CurPhase(Self, InLine(Self)) := ready
      InMess(Self) := noMess
  endif

transition R10 ==
 if (InMess(Self) = nackc and Home(InLine(Self)) = Self)
 then
      Pending(InLine(Self)) := false
      InMess(Self) := noMess
  endif
```
(*--*)

```
transition R9_R10 ==
  if InMess(Self) = nackc2
  then
      CurPhase(Self, InLine(Self)) := ready
      Pending(InLine(Self)) := false
      InMess(Self) := noMess
  endif
```

(*--*)

```
transition R11 ==
  if (InMess(Self) = getx and Home(InLine(Self)) = Self)
  then
      if Pending(InLine(Self)) = true
      then if MessInTr(n(maxQ),InSource(Self)) = noMess
          then AppendToTransit (InSource(Self),
                      (Self,nack,InSource(Self),InLine(Self)))
              InMess(Self):=noMess
          endif
      else if Owner(InLine(Self)) != none
          then if MessInTr(n(maxQ),Owner(InLine(Self))) = noMess
              then AppendToTransit (Owner(InLine(Self)),
                  (Self,fwdgetx,InSource(Self),InLine(Self)))
                  Pending(InLine(Self)) := true
                  InMess(Self):=noMess
              endif
          else if ( forall agent_ in Agent_without_none :
                              not(Sharer(InLine(Self),agent_)) )
              then if MessInTr(n(maxQ),InSource(Self)) = noMess
                  then AppendToTransit (InSource(Self),
                      (Self,putx,InSource(Self),InLine(Self)))
                      Owner(InLine(Self)) := InSource(Self)
                      InMess(Self):=noMess
                  endif
              else if (forall agent_ in Agent_without_none :
                              not(Sharer(InLine(Self),agent_))
                      or MessInTr(n(maxQ),agent_) = noMess )
                  then do forall agent_ in Agent_without_none
                              with Sharer(InLine(Self),agent_)
                      AppendToTransit (agent_,
                        (Self,inv,InSource(Self),InLine(Self)))
                      Pending(InLine(Self)):=true
                      enddo
                      InMess(Self):=noMess
                  endif
              endif
          endif
      endif
  endif
```

(*--*)

```
transition R12 ==
  if InMess(Self) = fwdgetx
  then
      if CCState(Self,InLine(Self)) = exclusive
      then if (Home(InLine(Self))=InSource(Self))
          then if MessInTr(n(maxQ),Home(InLine(Self))) = noMess
              then AppendToTransit (Home(InLine(Self)),
                    (Self,putx_fwdack,InSource(Self),InLine(Self)))
                  CCState(Self,InLine(Self)):=invalid
                  InMess(Self):=noMess
              endif
          else if MessInTr(n(maxQ),InSource(Self)) = noMess
                and MessInTr(n(maxQ),Home(InLine(Self))) = noMess
              then AppendToTransit (InSource(Self),
                        (Self,putx,InSource(Self),InLine(Self)))
                  AppendToTransit (Home(InLine(Self)),
                        (Self,fwdack,InSource(Self),InLine(Self)))
                  CCState(Self,InLine(Self)):=invalid
                  InMess(Self):=noMess
              endif
          endif
      else if (Home(InLine(Self))=InSource(Self))
          then if MessInTr(n(maxQ),Home(InLine(Self))) = noMess
              then AppendToTransit (InSource(Self),
                        (Self,nackc2,InSource(Self),InLine(Self)))
                  InMess(Self):=noMess
              endif
          else if MessInTr(n(maxQ),InSource(Self)) = noMess
                and MessInTr(n(maxQ),Home(InLine(Self))) = noMess
              then AppendToTransit (InSource(Self),
                        (Self,nack,InSource(Self),InLine(Self)))
                  AppendToTransit (Home(InLine(Self)),
                        (Self,nackc,InSource(Self),InLine(Self)))
                  InMess(Self):=noMess
              endif
          endif
      endif
  endif


transition R13 ==
  if InMess(Self) = putx
  then
      CCState(Self,InLine(Self)) := exclusive
      CurPhase(Self, InLine(Self)) := ready
      InMess(Self) := noMess
  endif

transition R14 ==
  if (InMess(Self) = fwdack and Home(InLine(Self)) = Self)
  then
      Owner(InLine(Self)) := InSource(Self)
      Pending(InLine(Self)) := false
```

153

```
      InMess(Self) := noMess
  endif


(*----------------------------------------------------------------*)

transition R13_R14 ==
  if InMess(Self) = putx_fwdack
  then
      CCState(Self,InLine(Self)) := exclusive
      CurPhase(Self, InLine(Self)) := ready
      Owner(InLine(Self)) := InSource(Self)
      Pending(InLine(Self)) := false
      InMess(Self) := noMess
  endif


(*----------------------------------------------------------------*)

transition R15 ==
  if InMess(Self) = inv
  then
      if MessInTr(n(maxQ),Home(InLine(Self))) = noMess
      then AppendToTransit (Home(InLine(Self)),
                  (Self,invack,InSource(Self),InLine(Self)))
           InMess(Self):=noMess
           if CCState(Self,InLine(Self)) = shared
           then CCState(Self,InLine(Self)) := invalid
           else if CurPhase(Self, InLine(Self)) = wait
                then CurPhase(Self, InLine(Self)) := invalidPhase
                endif
           endif
      endif
  endif


transition R16 ==
  if (InMess(Self) = invack and Home(InLine(Self)) = Self)
  then do forall agent_ in Agent_without_none with true
        if InSender(Self)=agent_
        then Sharer(InLine(Self),agent_) := false
            if ( forall other_agent_ in (Agent_without_none\{agent_})
                 : Sharer(InLine(Self), other_agent_)=false )
            then if MessInTr(n(maxQ),InSource(Self)) = noMess
                 then AppendToTransit (InSource(Self),
                             (Self,putx,InSource(Self),InLine(Self)))
                      Pending(InLine(Self)):=false
                      Owner(InLine(Self)) := InSource(Self)
                      InMess(Self):=noMess
                 endif
            else InMess(Self):=noMess
            endif
         endif
        enddo
```

154

```
    endif


(*----------------------------------------------------------------*)

transition R17 ==
 if (InMess(Self) = rpl and Home(InLine(Self)) = Self)
 then if Sharer(InLine(Self),InSender(Self)) = true
                 and Pending(InLine(Self)) = false
       then Sharer(InLine(Self),InSender(Self)) := false
             CCState(Self,InLine(Self)) := invalid
       endif
       InMess(Self) := noMess
   endif

transition R18 ==
 if (InMess(Self) = wb and Home(InLine(Self)) = Self)
 then if Owner(InLine(Self)) != none
       then Owner(InLine(Self)) := none
             CCState(Self,InLine(Self)) := invalid
       endif
       InMess(Self) := noMess
   endif

(*----------------------------------------------------------------*)

transition behavior ==
   block
     R_Requests
     R5
     R6
     R7
     R8
     R7_R8
     R9
     R10
     R9_R10
     R11
     R12
     R13
     R14
     R13_R14
     R15
     R16
     R17
     R18
   endblock



(*----------------------------------------------------------------*)

transition  ClearMessInTr(agent_) ==
block
```

```
  MessInTr(n(maxQ),agent_):=noMess
  if MessInTr(n(2),agent_)=noMess
  then MessInTr(n(1),agent_):=noMess
  else do forall i in {2..maxQ}
         if MessInTr(n(i), agent_)!=noMess
         then MessInTr (n(i-1), agent_):=MessInTr(n(i), agent_)
              SenderInTr (n(i-1), agent_):=SenderInTr(n(i), agent_)
              SourceInTr (n(i-1), agent_):=SourceInTr(n(i), agent_)
              LineInTr (n(i-1), agent_):=LineInTr(n(i), agent_)
         else MessInTr(n(i-1), agent_):=noMess
         endif
         enddo
  endif
endblock

transition SendMess(agent_) ==
  block
   InSender(agent_):= SenderInTr(n(1), agent_)
   InMess(agent_)   := MessInTr(n(1), agent_)
   InSource(agent_):= SourceInTr(n(1), agent_)
   InLine(agent_)   := LineInTr(n(1), agent_)
   ClearMessInTr(agent_)
   endblock


transition SendR(agent_) ==
  block
   InSender(agent_)   := SenderInTrR(agent_)
   InMess(agent_)     := MessInTrR (agent_)
   InSource(agent_)   := SourceInTrR (agent_)
   InLine(agent_)     := LineInTrR (agent_)
   MessInTrR(agent_)  := noMess
   endblock


transition SendRequest (agent_) ==
  if MessInTrR (agent_) = get
     and CurPhase(SenderInTrR(agent_), LineInTrR (agent_)) = ready
     and CCState(SenderInTrR(agent_), LineInTrR (agent_)) = invalid
  then SendR(agent_)
       CurPhase(SenderInTrR(agent_), LineInTrR (agent_)) := wait
  else if MessInTrR (agent_) = getx
          and CurPhase(SenderInTrR(agent_),LineInTrR(agent_))=ready
          and CCState(SenderInTrR(agent_),LineInTrR(agent_))=invalid
       then SendR(agent_)
            CurPhase(SenderInTrR(agent_),LineInTrR (agent_)) := wait
       else if MessInTrR (agent_) = rpl
               and CurPhase(SenderInTrR(agent_),
                              LineInTrR (agent_)) = ready
               and CCState(SenderInTrR(agent_),
                              LineInTrR(agent_)) = shared
            then SendR(agent_)
                 CCState(SenderInTrR(agent_),
```

```
                              LineInTrR(agent_)) := invalid
            else if MessInTrR (agent_) = wb
                    and CurPhase(SenderInTrR(agent_),
                                    LineInTrR(agent_)) = ready
                    and CCState(SenderInTrR(agent_),
                                    LineInTrR(agent_)) = exclusive
                    then SendR(agent_)
                        CCState(SenderInTrR(agent_),
                                    LineInTrR (agent_)) := invalid
                    endif
            endif
        endif
    endif


(*-------------------------------------------------------------------*)

transition synchronize ==
 block
  do forall agent_ in Agent_without_none with true
    if InMess(agent_)=noMess
    then if MessInTr(n(1),agent_) != noMess
            then SendMess(agent_)
            else if MessInTrR(agent_) != noMess
                    and InMess(agent_)=noMess
                    then SendRequest(agent_)
                    endif
        endif
    endif
  enddo
 endblock

(*-------------------------------------------------------------------*)

transition main ==
    if toggle=behave
    then
        behavior
        toggle:=sync
    else
        synchronize
        toggle:=behave
    endif

(*-------------------------------------------------------------------*)
```

A.2.2   The Counterexamples

Counterexample 1

Scenario: agent_2 has shared access to line_1 when agent_1 ask for exclusive access (state 1.6). *home* (i.e. agent_1) sends an invalidation message to agent_2 (state 1.10 – state 1.11) that is responsed with an invalidation acknowledgement by agent_2. Now the exclusive access is granted to agent_1 by sending a putx-message but without noting the new owner of line_1 (state 1.14 – state 1.15). Although the state of agent_1 is exclusive with respect to line_1, agent_2, when asking for in the next step (state 1.17), gets exclusive access, too (state 1.19). Mutual exclusion for exclusive access is violated (state 1.16, state 1.20).

Checked requirement specification is safety:
```
AG (!(State(agent_1, line_1) = exclusive
    & State(agent_2, line_1)=exclusive))

-- specification AG (!(State(agent_1, line_1... is false
-- as demonstrated by the following execution sequence
```

```
state 1.1:
State(agent_1, line_1) = invalid
State(agent_2, line_1) = invalid
CurPhase(agent_1, line_1) = ready
CurPhase(agent_2, line_1) = ready
InLine(agent_1) = line_1
InLine(agent_2) = line_1
InMess(agent_1) = noMess
InMess(agent_2) = noMess
InSender(agent_1) = agent_2
InSender(agent_2) = agent_2
InSource(agent_1) = agent_2
InSource(agent_2) = agent_2
LineInTr(agent_1) = line_1
LineInTr(agent_2) = line_1
LineInTrR(agent_1) = line_1
MessInTr(agent_1) = noMess
MessInTr(agent_2) = noMess
MessInTrR(agent_1) = noMess
Owner(line_1) = none
Pending(line_1) = 0
Self = agent_2
SenderInTr(agent_1) = agent_2
SenderInTr(agent_2) = agent_2
SenderInTrR(agent_1) = agent_2
Sharer(line_1, agent_1) = 0
Sharer(line_1, agent_2) = 0
SourceInTr(agent_1) = agent_2
SourceInTr(agent_2) = agent_2
SourceInTrR(agent_1) = agent_1
produce(agent_1) = (ccwb, line_1)
produce(agent_2) = (ccwb, line_1)
toggle = behave

state 1.2:
toggle = sync

state 1.3:
produce(agent_2) = (ccget, line_1)
toggle = behave

state 1.4:
CurPhase(agent_2, line_1) = wait
MessInTrR(agent_1) = get
SourceInTrR(agent_1) = agent_2
produce(agent_2) = (ccwb, line_1)
```

```
toggle = sync

state 1.5:
InMess(agent_1) = get
MessInTrR(agent_1) = noMess
Self = agent_1
produce(agent_1) = (ccgetx, line_1)
toggle = behave

state 1.6:
CurPhase(agent_1, line_1) = wait
MessInTr(agent_2) = put
MessInTrR(agent_1) = getx
Self = agent_2
SenderInTr(agent_2) = agent_1
SenderInTrR(agent_1) = agent_1
Sharer(line_1, agent_2) = 1
SourceInTrR(agent_1) = agent_1
produce(agent_1) = (ccwb, line_1)
toggle = sync

state 1.7:
InMess(agent_1) = noMess
InMess(agent_2) = put
InSender(agent_2) = agent_1
MessInTr(agent_2) = noMess
toggle = behave

state 1.8:
State(agent_2, line_1) = shared
CurPhase(agent_2, line_1) = ready
InMess(agent_2) = noMess
toggle = sync

state 1.9:
InMess(agent_1) = getx
InSender(agent_1) = agent_1
InSource(agent_1) = agent_1
MessInTrR(agent_1) = noMess
Self = agent_1
toggle = behave

state 1.10:
MessInTr(agent_2) = inv
Pending(line_1) = 1
Self = agent_2
```

```
SourceInTr(agent_2) = agent_1            toggle = behave
toggle = sync
                                         state 1.16:
state 1.11:                              State(agent_1, line_1) = exclusive
InMess(agent_1) = noMess                 CurPhase(agent_1, line_1) = ready
InMess(agent_2) = inv                    InMess(agent_1) = noMess
InSource(agent_2) = agent_1              Self = agent_2
MessInTr(agent_2) = noMess               toggle = sync
produce(agent_2) = (ccgetx, line_1)
toggle = behave                          state 1.17:
                                         InMess(agent_1) = getx
state 1.12:                              InSender(agent_1) = agent_2
State(agent_2, line_1) = invalid         InSource(agent_1) = agent_2
CurPhase(agent_2, line_1) = wait         MessInTrR(agent_1) = noMess
MessInTr(agent_1) = invack               Self = agent_1
MessInTrR(agent_1) = getx                produce(agent_1) = (ccget, line_1)
SenderInTrR(agent_1) = agent_2           toggle = behave
SourceInTr(agent_1) = agent_1
SourceInTrR(agent_1) = agent_2           state 1.18:
produce(agent_2) = (ccwb, line_1)        CurPhase(agent_1, line_1) = wait
toggle = sync                            MessInTr(agent_2) = putx
                                         MessInTrR(agent_1) = get
state 1.13:                              Owner(line_1) = agent_2
InMess(agent_1) = invack                 Self = agent_2
InMess(agent_2) = noMess                 SenderInTrR(agent_1) = agent_1
InSender(agent_1) = agent_2              SourceInTr(agent_2) = agent_2
MessInTr(agent_1) = noMess               SourceInTr(agent_1) = agent_1
Self = agent_1                           produce(agent_1) = (ccwb, line_1)
toggle = behave                          toggle = sync

state 1.14:                              state 1.19:
MessInTr(agent_1) = putx                 InMess(agent_1) = noMess
Pending(line_1) = 0                       InMess(agent_2) = putx
Self = agent_2                           InSource(agent_2) = agent_2
SenderInTr(agent_1) = agent_1            MessInTr(agent_2) = noMess
Sharer(line_1, agent_1) = 0              MessInTrR(agent_1) = noMess
toggle = sync                            toggle = behave

state 1.15:                              state 1.20:
InMess(agent_1) = putx                   State(agent_2, line_1) = exclusive
InSender(agent_1) = agent_1              CurPhase(agent_2, line_1) = ready
MessInTr(agent_1) = noMess               InMess(agent_2) = noMess
Self = agent_1                           toggle = sync
```

```
resources used:
user time: 1910.14 s, system time: 0.51 s
BDD nodes allocated: 1755356
Bytes allocated: 33161216
BDD nodes representing transition relation: 832699 + 54
```

Counterexample 2

Scenario: The length of the queue of messages in transit equals one and two agents are involved in communication. Both agents, agent_1 and agent_2, are sharers of line_1 (state 1.6, state 1.12). agent_2 sends an invalidation acknowledgement to *home* (i.e. agent_1). Then agent_1 tries to send its inv-message but it is blocked because the queue is full (state 1.20 and following). As being blocked, agent_1 as the home of the line cannot process the message of agent_2 in order to free the queue. In this situation a deadlock occurs.

Checked requirement specification is liveness:
```
AG AF (curPhase(agent_1, line_1) = ready)
  & AG AF (curPhase(agent_2, line_1) = ready)

-- specification AG AF CurPhase(agent_1, line_1... is false
-- as demonstrated by the following execution sequence
```

```
state 1.1:                                state 1.5:
State(agent_1, line_1) = invalid          InMess(agent_1) = get
State(agent_2, line_1) = invalid          InSender(agent_1) = agent_1
CurPhase(agent_1, lines_1) = ready        InSource(agent_1) = agent_1
CurPhase(agent_2, line_1) = ready         MessInTrR(agent_1) = noMess
InLine(agent_1) = line_1                  Self = agent_1
InLine(agent_2)= line_1                   toggle = behave
InMess(agent_1) = noMess
InMess(agent_2)= noMess                   state 1.6:
InSender(agent_1) = agent_2               MessInTr(agent_1) = put
InSender(agent_2)= agent_2                Self = agent_2
InSource(agent_1)= agent_2                SenderInTr(agent_1) = agent_1
InSource(agent_2)= agent_2                Sharer(line_1, agent_1) = 1
LineInTr(agent_1) = line_1                SourceInTr(agent_1) = agent_1
LineInTr(agent_2)= line_1                 toggle = sync
LineInTrR(agent_1) = line_1
MessInTr(agent_1) = noMess                state 1.7:
MessInTr(agent_2)= noMess                 InMess(agent_1) = put
MessInTrR(agent_1) = noMess               MessInTr(agent_1) = noMess
Owner(lines_1) = none                     Self = agent_1
Pending(line_1) = 0                       toggle = behave
Self = agent_2
SenderInTr(agent_1) = agent_2             state 1.8:
SenderInTr(agent_2)= agent_2              State(agent_1, line_1) = shared
SenderInTrR(agent_1) = agent_2            CurPhase(agent_1, line_1) = ready
Sharer(line_1, agent_1) = 0               InMess(agent_1) = noMess
Sharer(line_1,agent_2) = 0                Self = agent_2
SourceInTr(agent_1) = agent_2             toggle = sync
SourceInTr(agent_2) = agent_2
SourceInTrR(agent_1) = agent_1            state 1.9:
produce(agent_1) = (ccwb, line_1)         produce(agent_2) = (ccget, line_1)
produce(agent_2) = (ccwb, line_1)         toggle = behave
toggle = behave
                                          state 1.10:
state 1.2:                                CurPhase(agent_2, line_1) = wait
toggle = sync                             MessInTrR(agent_1) = get
                                          SenderInTrR(agent_1) = agent_2
state 1.3:                                SourceInTrR(agent_1) = agent_2
Self = agent_1                            produce(agent_2) = (ccwb, line_1)
produce(agent_1) = (ccget, line_1)        toggle = sync
toggle = behave
                                          state 1.11:
state 1.4:                                InMess(agent_1) = get
CurPhase(agent_1, line_1) = wait          InSender(agent_1) = agent_2
MessInTrR(agent_1) = get                  InSource(agent_1) = agent_2
Self = agent_2                            MessInTrR(agent_1) = noMess
SenderInTrR(agent_1) = agent_1            Self = agent_1
produce(agent_1) = (ccwb, line_1)         produce(agent_1) = (ccgetx, line_1)
toggle = sync                             toggle = behave
```

```
state 1.12:
CurPhase(agent_1, line_1) = wait
MessInTr(agent_2) = put
MessInTrR(agent_1) = getx
Self = agent_2
SenderInTr(agent_2) = agent_1
SenderInTrR(agent_1) = agent_1
Sharer(line_1, agent_2) = 1
SourceInTrR(agent_1) = agent_1
produce(agent_1) = (ccwb, line_1)
toggle = sync

state 1.13:
InMess(agent_1) = noMess
InMess(agent_2) = put
InSender(agent_2) = agent_1
MessInTr(agent_2) = noMess
Self = agent_1
toggle = behave

state 1.14:
Self = agent_2
toggle = sync

state 1.15:
InMess(agent_1) = getx
InSender(agent_1) = agent_1
InSource(agent_1) = agent_1
MessInTrR(agent_1) = noMess
toggle = behave

state 1.16:
State(agent_2, line_1) = shared
CurPhase(agent_2, line_1) = ready
InMess(agent_2) = noMess
toggle = sync

state 1.17:
Self = agent_1
```

```
toggle = behave

state 1.18:
MessInTr(agent_1) = inv
MessInTr(agent_2) = inv
Pending(line_1) = 1
Self = agent_2
SourceInTr(agent_2) = agent_1
toggle = sync

state 1.19:
InMess(agent_1) = inv
InMess(agent_2) = inv
InSource(agent_2) = agent_1
MessInTr(agent_1) = noMess
MessInTr(agent_2) = noMess
toggle = behave

-- loop starts here --
state 1.20:
State(agent_2, line_1) = invalid
MessInTr(agent_1) = invack
SenderInTr(agent_1) = agent_2
toggle = sync

state 1.21:
Self = agent_1
toggle = behave

-- loop starts here --
state 1.22:
Self = agent_2
toggle = sync

state 1.23:
toggle = behave

state 1.24:
toggle = sync
```

```
resources used:
user time: 2391.1 s, system time: 0.64 s
BDD nodes allocated: 1754532
Bytes allocated: 33161216
BDD nodes representing transition relation: 658216 + 54
```

Counterexample 3

Scenario: The oracle produce(agent_1,line_1) produces a write-back request wb, the corresponding request is sent to the queue for request messages in transit MessInTrR (state 1.1). This request message is not passed through to the *home* but has to wait in the queue. In the meanwhile, **agent_1** generates a new request for exclusive access and gets the owner of the line (state 1.8). Then **agent_2** requests for exclusive access and as a consequence **agent_1** invalidates its status of exclusive access to line_1 and **agent_2** becomes owner of the line (state 1.14 and state 1.15). Then the late wb-request of **agent_1** is transfered and owner(line_1) is set to none (state 1.18), although **agent_2** was the current owner of to the line (state 1.16) and not **agent_1**. Now its possible for **agent_1** to get exclusive access to the line too (state 1.20, state 1.22).

Checked requirement specification is safety:
```
AG (!(State(agent_1, line_1) = exclusive
     & State(agent_2, line_1)=exclusive))

-- specification AG (!(State(agent_1, line_1... is false
-- as demonstrated by the following execution sequence
```

```
state 1.1:
State(agent_1, line_1) = invalid
State(agent_2, line_1) = invalid
CurPhase(agent_1, line_1) = ready
CurPhase(agent_2, line_1) = ready
InLine(agent_1) = line_1
InLine(agent_2) = line_1
InMess(agent_1) = noMess
InMess(agent_2) = noMess
InSender(agent_1) = agent_2
InSender(agent_2) = agent_2
InSource(agent_1) = agent_2
InSource(agent_2) = agent_2
LineInTr(agent_1) = line_1
LineInTr(agent_2) = line_1
LineInTrR(agent_1) = line_1
MessInTr(agent_1) = noMess
MessInTr(agent_2) = noMess
MessInTrR(agent_1) = noMess
Owner(line_1) = none
Pending(line_1) = 0
Self = agent_2
SenderInTr(agent_1) = agent_2
SenderInTr(agent_2) = agent_2
SenderInTrR(agent_1) = agent_2
Sharer(line_1, agent_1) = 0
Sharer(line_1, agent_2) = 0
SourceInTr(agent_1) = agent_2
SourceInTr(agent_2) = agent_2
SourceInTrR(agent_1) = agent_1
produce(agent_1) = (ccwb, line_1)
produce(agent_2) = (ccwb, line_1)
toggle = behave

state 1.2:
toggle = sync

state 1.3:
Self = agent_1
produce(agent_1) = (ccgetx, line_1)
toggle = behave

state 1.4:
CurPhase(agent_1, line_1) = wait
MessInTrR(agent_1) = getx
Self = agent_2
SenderInTrR(agent_1) = agent_1
produce(agent_1) = (ccwb, line_1)
toggle = sync
```

```
state 1.5:
InMess(agent_1) = getx
InSender(agent_1) = agent_1
InSource(agent_1) = agent_1
MessInTrR(agent_1) = noMess
Self = agent_1
toggle = behave

state 1.6:
MessInTr(agent_1) = putx
Owner(line_1) = agent_1
Self = agent_2
SenderInTr(agent_1) = agent_1
SourceInTr(agent_1) = agent_1
toggle = sync

state 1.7:
InMess(agent_1) = putx
MessInTr(agent_1) = noMess
Self = agent_1
toggle = behave

state 1.8:
State(agent_1, line_1) = exclusive
CurPhase(agent_1, line_1) = ready
InMess(agent_1) = noMess
Self = agent_2
toggle = sync

state 1.9:
produce(agent_2) = (ccgetx, line_1)
toggle = behave

state 1.10:
CurPhase(agent_2, line_1) = wait
MessInTrR(agent_1) = getx
SenderInTrR(agent_1) = agent_2
SourceInTrR(agent_1) = agent_2
produce(agent_2) = (ccwb, line_1)
toggle = sync

state 1.11:
InMess(agent_1) = getx
InSender(agent_1) = agent_2
InSource(agent_1) = agent_2
MessInTrR(agent_1) = noMess
Self = agent_1
produce(agent_1) = undef
toggle = behave
```

state 1.12:
MessInTr(agent_1) = fwdgetx
Pending(line_1) = 1
Self = agent_2
SourceInTr(agent_1) = agent_2
produce(agent_1) = (ccwb, line_1)
toggle = sync

state 1.13:
InMess(agent_1) = fwdgetx
InSender(agent_1) = agent_1
MessInTr(agent_1) = noMess
Self = agent_1
toggle = behave

state 1.14:
State(agent_1, line_1) = invalid
MessInTr(agent_1) = fwdack
MessInTr(agent_2) = putx
MessInTrR(agent_1) = wb
Self = agent_2
SenderInTr(agent_2) = agent_1
SenderInTrR(agent_1) = agent_1
SourceInTrR(agent_1) = agent_1
toggle = sync

state 1.15:
InMess(agent_1) = fwdack
InMess(agent_2) = putx
InSender(agent_2) = agent_1
MessInTr(agent_1) = noMess
MessInTr(agent_2) = noMess
Self = agent_1
toggle = behave

state 1.16:
InMess(agent_1) = noMess
Owner(line_1) = agent_2
Pending(line_1) = 0
Self = agent_2
toggle = sync

state 1.17:
InMess(agent_1) = wb
InSource(agent_1) = agent_1
MessInTrR(agent_1) = noMess

Self = agent_1
produce(agent_1) = (ccgetx, line_1)
toggle = behave

state 1.18:
CurPhase(agent_1, line_1) = wait
InMess(agent_1) = noMess
MessInTrR(agent_1) = getx
Owne(line_1) = none
Self = agent_2
produce(agent_1) = (ccwb, line_1)
toggle = sync

state 1.19:
InMess(agent_1) = getx
MessInTrR(agent_1) = noMess
Self = agent_1
toggle = behave

state 1.20:
MessInTr(agent_1) = putx
Owner(line_1) = agent_1
Self = agent_2
SourceInTr(agent_1) = agent_1
toggle = sync

state 1.21:
InMess(agent_1) = putx
MessInTr(agent_1) = noMess
Self = agent_1
toggle = behave

state 1.22:
State(agent_1, line_1) = exclusive
CurPhase(agent_1, line_1) = ready
InMess(agent_1) = noMess
Self = agent_2
toggle = sync

state 1.23:
toggle = behave

state 1.24:
State(agent_2, line_1) = exclusive
CurPhase(agent_2, line_1) = ready
InMess(agent_2) = noMess
toggle = sync

resources used:
user time: 1244.76 s, system time: 0.4 s
BDD nodes allocated: 1754532
Bytes allocated: 33161216
BDD nodes representing transition relation: 658216 + 54

A.3 Production Cell

A.3.1 The ASM-SL Code

```
(*----------------------------------------------------------------*)
(* Datatypes *)

datatype SWITCH_1 == {on, off}
datatype SWITCH_2 == {up, down, idle}
datatype SWITCH_3 == {clockwise, counterClockwise, idle2}
datatype SWITCH_4 == {extending, retract, idle3}
datatype SWITCH_5 == {run, idle4}
datatype SWITCH_6 == {toFeedBelt, toDepBelt, idle5}

datatype SENSOR_1 == {Arm1IntoPress, OverTable, retracted,
                      Arm2IntoPress, OverDepBelt}
datatype SENSOR_2 == {Arm1ToTable, Arm2ToPress, Arm2ToDepBelt,
                      Arm1ToPress}
datatype SENSOR_3 == {OnDepBelt, OnFeedBelt, SafeDistanceFromFeedBelt}


(*----------------------------------------------------------------*)
(* Dynamic functions *)

(* Module FeedBelt: *)

dynamic function FeedBeltMot :SWITCH_1
        initially on

dynamic function Delivering :BOOL
        initially false


dynamic function FeedBeltFree :BOOL
        initially true


(*Module Table *)

dynamic function TableLoaded :BOOL
        initially false

dynamic function TableElevationMot :SWITCH_2
        initially idle

dynamic function TableRotationMot :SWITCH_3
        initially idle2


(* MODULE Robot *)

dynamic function Arm1Mot :SWITCH_4
```

```
        initially idle3

dynamic function Arm2Mot :SWITCH_4
        initially idle3

dynamic function RobotRotationMot :SWITCH_3
        initially idle2

dynamic function Arm1Mag :SWITCH_1
        initially off

dynamic function Arm2Mag :SWITCH_1
        initially off

(* Module Press *)

dynamic function PressMot :SWITCH_2
        initially idle

dynamic function PressLoaded :BOOL
        initially true

(* Module DepositBelt *)

dynamic function DepBeltMot :SWITCH_5
        initially run

dynamic function Critical :BOOL
        initially false

dynamic function PieceAtDepositBeltEnd :BOOL
        initially false

dynamic function DepositBeltReadyForLoading :BOOL
        initially true

(* Module Crane *)

dynamic function CraneHorizontalMot :SWITCH_6
        initially idle5

dynamic function CraneVerticalMot :SWITCH_2
        initially idle

dynamic function CraneMagnet :SWITCH_1
        initially off

(*----------------------------------*)
(* Sensorvariables or Orakelfunctions *)

external function PieceInFeedBeltLightBarrier :BOOL
external function MaxRotation :BOOL
external function MinRotation :BOOL
external function TopPosition :BOOL
```

```
external function BottomPosition :BOOL
external function Arm1Ext :SENSOR_1
external function Arm2Ext :SENSOR_1
external function Angle :SENSOR_2
external function TopPositionPress :BOOL
external function MiddlePositionPress :BOOL
external function BottomPositionPress :BOOL
external function ForgingComplete :BOOL
external function PieceInDepositBeltLightBarrier :BOOL
external function GripperOverDepBelt :BOOL
external function GripperOverFeedBelt :BOOL
external function GripperVerticalPos :SENSOR_3

(*----------------------------------------------------------*)
(* MODULE FeedBelt                                          *)

transition FB_NORMAL ==
      if (FeedBeltMot = on) and not(Delivering) and
         (PieceInFeedBeltLightBarrier = true)
      then
        block
          FeedBeltFree := true
          if ((BottomPosition=true) and (MinRotation=true)
               and (TableElevationMot=idle)
               and (TableRotationMot=idle2))
            and not(TableLoaded)
             then Delivering := true
             else FeedBeltMot := off
          endif
        endblock
      endif


transition FB_STOPPED ==
      if (FeedBeltMot = off)
        and ((BottomPosition=true) and (MinRotation=true)
        and (TableElevationMot=idle) and (TableRotationMot=idle2))
        and not(TableLoaded)
      then FeedBeltMot := on
           Delivering := true
      endif


transition FB_CRITICAL ==
      if (FeedBeltMot = on) and Delivering
        and not(PieceInFeedBeltLightBarrier)
      then Delivering := false
           TableLoaded := true
      endif



(*----------------------------------------------------------*)
(* MODULE Table                                             *)
```

```
transition WAITING_LOAD ==
        if ((BottomPosition = true) and (MinRotation = true)
          and (TableElevationMot = idle)
          and (TableRotationMot = idle2) and (TableLoaded = true)
        then TableElevationMot := up
             TableRotationMot  := clockwise
        endif


transition MOVING_UNLOAD_A ==
        if (TableElevationMot = up) and (TopPosition = true)
        then TableElevationMot := idle
        endif

transition MOVING_UNLOAD_B ==
        if (TableRotationMot = clockwise) and (MaxRotation = true)
        then TableRotationMot := idle2
        endif

transition WAITING_UNLOAD ==
        if ((TopPosition = true) and (MaxRotation = true)
            and (TableElevationMot = idle)
            and (TableRotationMot = idle2))
          and (TableLoaded = true)
        then TableElevationMot := down
             TableRotationMot  := counterClockwise
        endif

transition MOVING_LOAD_A ==
        if (TableElevationMot = down) and (BottomPosition = true)
        then TableElevationMot := idle
        endif

transition MOVING_LOAD_B ==
        if (TableRotationMot = counterClockwise)
          and (MinRotation = true)
        then TableRotationMot := idle2
        endif

(*----------------------------------------------------------*)
(* MODULE Robot                                             *)

transition WAITING1 ==
   if (Angle=Arm1ToTable) and (Arm1Ext=retracted)
     and (Arm2Ext=retracted) and (RobotRotationMot=idle2)
     and (Arm1Mot=idle3) and (Arm2Mot=idle3) and (Arm1Mag=off)
     and (Arm2Mag=off) and TopPosition and MaxRotation
     and TableLoaded
   then Arm1Mot:=extending
   endif

transition WAITING2 ==
   if(Angle=Arm2ToPress) and (Arm1Ext=retracted)
```

```
      and (Arm2Ext=retracted) and (RobotRotationMot=idle2)
      and (Arm1Mot=idle3) and (Arm2Mot=idle3) and (Arm1Mag=on)
      and (Arm2Mag=off) and BottomPosition and (PressMot=idle)
      and PressLoaded
    then Arm2Mot:=extending
    endif

transition WAITING3 ==
    if (Angle=Arm2ToDepBelt) and (Arm1Ext=retracted)
      and (Arm2Ext=retracted) and (RobotRotationMot=idle2)
      and (Arm1Mot=idle3) and (Arm2Mot=idle3) and (Arm1Mag=on)
      and (Arm2Mag=on) and DepositBeltReadyForLoading
    then Arm2Mot:=extending
    endif


transition WAITING4 ==
    if (Angle=Arm2ToDepBelt) and (Arm1Ext=retracted)
      and (Arm2Ext=retracted) and (RobotRotationMot=idle2)
      and (Arm1Mot=idle3) and (Arm2Mot=idle3) and (Arm1Mag=on)
      and (Arm2Mag=on) and  MiddlePositionPress and (PressMot=idle)
      and not(PressLoaded)
    then Arm1Mot:=extending
    endif

transition ACTION_extension1 ==
    if (Angle=Arm1ToTable) and (Arm1Mot=extending)
    and (Arm1Ext=OverTable)
    then
      block
       Arm1Mot:=idle3
       Arm1Mag:=on
      endblock
    endif

transition ACTION_extension2 ==
    if (Angle=Arm2ToPress) and (Arm2Mot=extending)
    and (Arm2Ext=Arm2IntoPress)
    then
      block
       Arm2Mot:=idle3
       Arm2Mag:=on
      endblock
    endif

transition ACTION_extension3 ==
    if (Angle=Arm2ToDepBelt) and (Arm2Mot=extending)
    and (Arm2Ext=OverDepBelt)
    then
      block
       Arm2Mot:=idle3
       Arm2Mag:=off
      endblock
    endif
```

168

```
transition ACTION_extension4 ==
    if (Angle=Arm1ToPress) and (Arm1Mot=extending)
    and (Arm1Ext=Arm1IntoPress)
    then
      block
        Arm1Mot:=idle3
        Arm1Mag:=off
      endblock
    endif

(*-------------------------*)

transition ACTION_proper1 ==
    if (Angle=Arm1ToTable) and (Arm1Ext=OverTable)
       and (Arm1Mot=idle3) and (Arm1Mag=on)
    then
      Arm1Mot:=retract
    endif

transition ACTION_proper2 ==
    if (Angle=Arm2ToPress) and (Arm2Ext=Arm2IntoPress)
       and (Arm2Mot=idle3) and (Arm2Mag=on)
    then
      Arm2Mot:=retract
    endif

transition ACTION_proper3 ==
    if (Angle=Arm2ToDepBelt) and (Arm2Ext=OverDepBelt)
       and (Arm2Mot=idle3) and (Arm2Mag=off)
    then
      Arm2Mot:=retract
    endif

transition ACTION_proper4 ==
    if (Angle=Arm1ToPress) and (Arm1Ext=Arm1IntoPress)
       and (Arm1Mot=idle3) and (Arm1Mag=off)
    then
      Arm1Mot:=retract
    endif

(*-------------------------*)

transition ACTION_retraction1 ==
    if (Angle=Arm1ToTable) and (Arm1Mot=retract)
                                    and (Arm1Ext=retracted)
    then
     block
       RobotRotationMot:=counterClockwise
       TableLoaded:=false
     endblock
    endif

transition ACTION_retraction2 ==
```

```
  if (Angle=Arm2ToPress) and (Arm2Mot=retract)
                                      and (Arm2Ext=retracted)
  then
   block
     RobotRotationMot:=counterClockwise
     PressLoaded:=false
   endblock
  endif

transition ACTION_retraction3 ==
  if (Angle=Arm2ToDepBelt) and (Arm2Mot=retract)
                                      and (Arm2Ext=retracted)
  then
   block
     RobotRotationMot:=counterClockwise
     DepositBeltReadyForLoading:=false
   endblock
  endif

transition ACTION_retraction4 ==
  if (Angle=Arm1ToPress) and (Arm1Mot=retract)
                                      and (Arm1Ext=retracted)
  then
   block
     RobotRotationMot:=clockwise
     PressLoaded:=true
   endblock
  endif

(*--------------------------*)

transition MOVING1 ==
  if (Arm1Ext=retracted) and (Arm2Ext=retracted)
     and (Arm1Mot=idle3) and (Arm2Mot=idle3)
     and (RobotRotationMot=counterClockwise)
     and (Arm1Mag=on) and (Arm2Mag=off)
     and (Angle=Arm2ToPress)
  then
     RobotRotationMot:=idle2
  endif

transition MOVING2 ==
  if (Arm1Ext=retracted) and (Arm2Ext=retracted)
     and (Arm1Mot=idle3) and (Arm2Mot=idle3)
     and (RobotRotationMot=counterClockwise)
     and (Arm1Mag=on) and (Arm2Mag=on)
     and (Angle=Arm2ToDepBelt)
  then
     RobotRotationMot:=idle2
  endif

transition MOVING3 ==
  if (Arm1Ext=retracted) and (Arm2Ext=retracted)
```

```
      and (Arm1Mot=idle3) and (Arm2Mot=idle3)
      and (RobotRotationMot=counterClockwise)
      and (Arm1Mag=on) and (Arm2Mag=off)
      and (Angle=Arm1ToPress)
   then
      RobotRotationMot:=idle2
   endif

transition MOVING4 ==
   if (Arm1Ext=retracted) and (Arm2Ext=retracted)
      and (Arm1Mot=idle3) and (Arm2Mot=idle3)
      and (RobotRotationMot=clockwise)
      and (Arm1Mag=off) and (Arm2Mag=off)
      and (Angle=Arm1ToTable)
   then
      RobotRotationMot:=idle2
   endif


(*-------------------------------------------------------*)
(* Module Press                                       *)

transition WAITING_UNLOAD_PRESS ==
   if BottomPositionPress and (PressMot=idle)
                                 and (PressLoaded=false)
   then
      PressMot:=up
   endif

transition MOVING_TO_MIDDLE_PRESS ==
   if not(PressLoaded) and (PressMot=up)
                                 and MiddlePositionPress
   then
      PressMot:=idle
   endif

transition WAITING_LOAD_PRESS ==
   if MiddlePositionPress and (PressMot=idle)
                                    and PressLoaded
   then
      PressMot:=up
   endif

transition MOVING_TO_UPPER_PRESS ==
   if PressLoaded and (PressMot=up)
   then
      PressMot:=idle
   endif

transition CLOSED_PRESS ==
   if TopPositionPress and (PressMot=idle)
   then
      PressMot:=down
   endif
```

```
transition MOVING_TO_LOWER_PRESS ==
  if (PressMot=down) and BottomPositionPress
  then
    PressMot:=idle
  endif


(*----------------------------------------------------------*)
(* Module Deposit Belt                                      *)

transition DB_NORMAL ==
  if (DepBeltMot=run) and not(Critical)
                             and PieceInDepositBeltLightBarrier
  then
    Critical:=true
  endif

transition DB_CRITICAL ==
  if (DepBeltMot=run) and Critical
                        and not(PieceInDepositBeltLightBarrier)
  then
    block
      DepBeltMot:=idle4
      Critical:=false
      PieceAtDepositBeltEnd:=true
    endblock
  endif

transition DB_STOPPED ==
  if (DepBeltMot=idle4) and not(PieceAtDepositBeltEnd)
  then
    DepBeltMot:=run
  endif


(*----------------------------------------------------------*)
(* Module Crane                                             *)

transition WAITING_DB ==
  if GripperOverDepBelt and (GripperVerticalPos=OnDepBelt) and
  (CraneHorizontalMot=idle5) and (CraneVerticalMot=idle) and
  (CraneMagnet=off)
  and PieceAtDepositBeltEnd
  then
    CraneMagnet:=on
  endif


transition UNLOADING_DBa ==
  if (CraneVerticalMot=idle) and (GripperVerticalPos=OnDepBelt)
  and (CraneMagnet=on)
  then
    CraneVerticalMot:=up
```

```
    endif

transition UNLOADING_DBb ==
  if (CraneVerticalMot=up)
              and (GripperVerticalPos=SafeDistanceFromFeedBelt)
  then
    block
      CraneVerticalMot:=idle
      CraneHorizontalMot:=toFeedBelt
      DepositBeltReadyForLoading:=true
      PieceAtDepositBeltEnd:=false
    endblock
  endif


transition MOVING_FB ==
  if (CraneHorizontalMot=toFeedBelt) and GripperOverFeedBelt
  then
    CraneHorizontalMot:=idle5
  endif

transition WAITING_FB ==
  if GripperOverFeedBelt
     and (GripperVerticalPos=SafeDistanceFromFeedBelt)
     and (CraneHorizontalMot=idle5) and (CraneVerticalMot=idle)
     and (CraneMagnet=on) and FeedBeltFree
  then
    CraneVerticalMot:=down
  endif

transition LOADING_FBa ==
  if (CraneVerticalMot=down) and (GripperVerticalPos=OnFeedBelt)
     and GripperOverFeedBelt
  then
    block
      CraneVerticalMot:=idle
      CraneMagnet:=off
    endblock
  endif

transition LOADING_FBb ==
  if (CraneVerticalMot=idle) and (GripperVerticalPos=OnFeedBelt)
     and GripperOverFeedBelt and (CraneHorizontalMot=idle5)
     and (CraneMagnet=off)
  then
    block
      CraneHorizontalMot:=toDepBelt
      FeedBeltFree:=false
    endblock
  endif


transition MOVING_DBa ==
  if (CraneHorizontalMot=toDepBelt) and GripperOverDepBelt
```

```
  then
    block
      CraneHorizontalMot:=idle5
      CraneVerticalMot:=down
    endblock
  endif

transition MOVING_DBb ==
  if GripperOverDepBelt and (CraneVerticalMot=down)
     and (GripperVerticalPos=OnDepBelt)
  then
      CraneVerticalMot:=idle
  endif

(*----------------------------------------------------------------*)

transition productionCell ==
  block
        FB_NORMAL
        FB_STOPPED
        FB_CRITICAL
        WAITING_LOAD
        MOVING_UNLOAD_A
        MOVING_UNLOAD_B
        WAITING_UNLOAD
        MOVING_LOAD_A
        MOVING_LOAD_B
        WAITING1
        WAITING2
        WAITING3
        WAITING4
        ACTION_extension1
        ACTION_extension2
        ACTION_extension3
        ACTION_extension4
        ACTION_proper1
        ACTION_proper2
        ACTION_proper3
        ACTION_proper4
        ACTION_retraction1
        ACTION_retraction2
        ACTION_retraction3
        ACTION_retraction4
        MOVING1 MOVING2
        MOVING3 MOVING4
        WAITING_UNLOAD_PRESS
        MOVING_TO_MIDDLE_PRESS
        WAITING_LOAD_PRESS
        MOVING_TO_UPPER_PRESS
        CLOSED_PRESS
        MOVING_TO_LOWER_PRESS
        DB_NORMAL DB_CRITICAL
        DB_STOPPED WAITING_DB
        UNLOADING_DBa
```

```
      UNLOADING_DBb
      MOVING_FB
      WAITING_FB
      LOADING_FBa
      LOADING_FBb
      MOVING_DBa
      MOVING_DBb
   endblock
```

(*--*)

A.3.2   The Environment Model

When modelling the behaviour of the environment, we add the following transition
rule to the system model. Additionally, we have to change the external functions
into dynamic functions. In order to express that a mechanical device should halt in
a position for at least one step, we also introduce three status functions (see below).

(*--*)

```
datatype AngleStatus   == { arm1stoppedattable, arm2stoppedatpress,
                            arm2stoppedatbelt, arm1stoppedatpress }
datatype PressStatus   == { pressattop, pressatmid, pressatbot }
datatype GripperStatus == { gripperstoppedat_ovb, gripperstoppedat_ovf,
                            gripperstoppedat_sd}

dynamic function angle_status : AngleStatus
dynamic function press_status : PressStatus
dynamic function gripper_status : GripperStatus
```

(*--*)

```
transition environment ==
 block
   if (FeedBeltMot=on) and not(FeedBeltFree) and not(Delivering)
   then
       PieceInFeedBeltLightBarrier := true
   else if (FeedBeltMot=on) and Delivering
        then
             PieceInFeedBeltLightBarrier := false
        endif
   endif
   (*------------------------------------*)
   if (TableElevationMot=up)
   then
       TopPosition := true
       BottomPosition := false
   else if (TableElevationMot=down)
        then
             TopPosition := false
             BottomPosition := true
        endif
   endif
   (*------------------------------------*)
```

175

```
if (TableRotationMot=clockwise)
then
    MaxRotation := true
    MinRotation := false
else if (TableRotationMot=counterClockwise)
     then
         MaxRotation := false
         MinRotation := true
     endif
 endif
(*-------------------------------------*)
 if (Arm1Mot=extending) and (Angle=Arm1ToTable)
 then
      Arm1Ext := OverTable
 else if (Arm1Mot=extending) and (Angle=Arm1ToPress)
      then
           Arm1Ext := Arm1IntoPress
      else if (Arm1Mot=retract)
           then
                Arm1Ext := retracted
           endif
      endif
 endif
(*-------------------------------------*)
 if (Arm2Mot=extending) and (Angle=Arm2ToPress)
 then
      Arm2Ext := Arm2IntoPress
 else if (Arm2Mot=extending) and (Angle=Arm2ToDeptBelt)
      then
           Arm2Ext := OverDepBelt
      else if (Arm2Mot=retract)
           then
                Arm2Ext := retracted
           endif
      endif
 endif
(*-------------------------------------*)
 if (Angle=arm1totable) and (RobotRotationMot=idle2)
 then
     angle_status := arm1stoppedattable
 endif

 if (Angle=Arm2ToPress) and (RobotRotationMot=idle2)
 then
     angle_status := arm2stoppedatpress
 endif

 if (Angle=Arm2ToDepBelt) and (RobotRotationMot=idle2)
 then
     angle_status := arm2stoppedatbelt
 endif

 if (Angle=Arm1ToPress) and (RobotRotationMot=idle2)
 then
```

```
      angle_status := arm1stoppedatpress
   endif
(*------------------------------------*)
  if (Angle=Arm1ToTable) and (RobotRotationMot=counterClockwise)
      and (angle_status=arm1stoppedattable)
  then
      Angle := Arm2ToPress
  endif

  if (Angle=Arm2ToPress) and (RobotRotationMot=counterClockwise)
      and (angle_status=arm2stoppedatpress)
  then
      Angle := Arm2ToDepBelt
  endif

  if (Angle=Arm2ToDepBelt) and (RobotRotationMot=counterClockwise)
      and (angle_status=arm2stoppedatbelt)
  then
      Angle := Arm1ToPress
  endif

  if (Angle=Arm1ToPress) and (RobotRotationMot=clockwise)
      and (angle_status=arm1stoppedatpress)
  then
      Angle := Arm1ToTable
  endif
(*------------------------------------*)
  if (PressMot=idle)
  then if (TopPositionPress=true)
      then
          press_status := pressattop
          if (PressLoaded=true)
          then
              ForgingComplete := true
          endif
      else if (MiddlePositionPress=true)
          then
              press_status := pressatmid
          else if (BottomPositionPress=true)
              then
                  press_status := pressatbot
              endif
          endif
      endif
  endif

  if (PressLoaded=false)
  then
      ForgingComplete := false
  endif
(*------------------------------------*)
  if (TopPositionPress=true) and (PressMot=down)
  then
      TopPositionPress := false
```

177

```
        BottomPositionPress := true
    endif

    if (MiddlePositionPress=true) and (PressMot=up)
        and (press_status=pressatmid)
    then
        TopPositionPress := true
        MiddlePositionPress := false
    endif

    if (BottomPositionPress=true) and (PressMot=up)
    then
        MiddlePositionPress := true
        BottomPositionPress := false
    endif
(*-------------------------------------*)
    if (DepBeltMot=run)
    then if (DepositBeltReadyForLoading=false)
            and (PieceInDepositBeltLightBarrier=false)
        then
            PieceInDepositBeltLightBarrier := true
        else if (Critical=true)
                and (PieceInDepositBeltLightBarrier=true)
            then
                PieceInDepositBeltLightBarrier := false
            endif
        endif
    endif
(*-------------------------------------*)
    if (CraneHorizontalMot=toDepBelt)
    then
        GripperOverDepBelt := true
        GripperOverFeedBelt := false
    endif

    if (CraneHorizontalMot=toFeedBelt)
    then
        GripperOverDepBelt := false
        GripperOverFeedBelt := true
    endif
(*-------------------------------------*)
    if (CraneVerticalMot=idle)
    then if (GripperVerticalPos=OnDepositBelt)
        then
            gripper_status := gripperstoppedat_ovb
        else if (GripperVerticalPos=OnFeedBelt)
            then
                gripper_status := gripperstoppedat_ovf
            else if (GripperVerticalPos=SafeDistanceFromFeedBelt)
                then
                    gripper_status := gripperstoppedat_sd
                endif
            endif
        endif
```

```
      endif
(*----------------------------------------*)
    if (GripperVerticalPos=OnFeedBelt) and (CraneVerticalMot=up)
        and (gripper_status=gripperstoppedat_ovb)
    then
        GripperVerticalPos := SafeDistanceFromFeedBelt
    endif

    if (GripperVerticalPos=SafeDistanceFromFeedBelt)
        and (CraneVerticalMot=down)
        and (gripper_status=gripperstoppedat_sd)
    then
        GripperVerticalPos := OnFeedBelt
    endif

    if (GripperVerticalPos=OnFeedBelt) and (CraneVerticalMot=down)
        and (gripper_status=gripperstoppedat_ovf)
    then
        GripperVerticalPos := OnDepositBelt
    endif

    if (GripperVerticalPos=OnDepositBelt) and (CraneVerticalMot=up)
        and (gripper_status=gripperstoppedat_ovf)
    then
        GripperVerticalPos := SafeDistanceFromFeedBelt
    endif
 endblock

(*-------------------------------------------------------------*)
```

A.3.3   The Safety Requirements

The safety requirements of the Production Cell are given as properties of its components, but the simplicity of the system allows us to check the system as a whole. We specify each property as a single formula and form the conjunction of this set of formulas in order to get the requirement specification. The formalisation of the requirements in CTL is given in the following.

Safety of the Feed Belt   The safety of the feed belt requires, that *the feed belt does not put metal blanks on the table if the latter is already loaded or not stopped in load position.*

We express this property in terms of the abstract ASM model as follows: The module *FeedBelt* is not in the critical section unless the table is ready for loading.

The critical section is when the motor of the feed belt is running and the boolean variable *Delivering* is true, and no blank is in the light barrier section of the belt anymore (since the blank is being moved from the belt to the table). This is specified by a conjunction of three conditions on state variables:

$$FeedBeltMot = on \ \land \ Delivering = true$$
$$\land \ \neg PieceInFeedBeltLightBarrier$$

The macro *TableReadyForLoading* in the abstract model substitutes a combined condition as follows: the table is not loaded ($\neg TableLoaded$), the table is in bottom position ($BottomPosition$), the rotation of the table is at its minimal position ($MinRotation$), and the elevation motor and the rotation motor of the table are idle ($TableElevationMot = idle$, $TableRotationMot = idle$). Therefore, the formalisation of the feed belt safety property has the following form:

$$\mathbf{AG}\ (FeedBeltMot = on\ \wedge\ Delivering = true$$
$$\wedge\ \neg PieceInFeedBeltLightBarrier$$
$$\Rightarrow \neg TableLoaded\ \wedge\ BottomPosition\ \wedge\ MinRotation\ \wedge$$
$$TableElevationMot = idle\ \wedge\ TableRotationMot = idle)$$

Safety of the Elevating Rotary Table   The safety property of the table postulates that *the table never moves over its bounds.* These bounds are represented by the top and the bottom positions, and by the minimal and maximal rotation.

This property can be modelled as a conjunction of the requirements that none of the four bounds is crossed. In the model of the cell, however, the fact that the table has reached a bound and the motor is on is a guard in the transition rule for switching the motor off. Therefore, we can only require that whenever a bound is reached the motor should be switched off in the following state. It should be pointed out that, for the real-life system, the sensor indicating that a bound is reached should react soon enough to assure safety for the table.

The following CTL formula models the safety requirement for the table.

$$\mathbf{AG}\ (TopPosition \Rightarrow \mathbf{AX}\ (TableElevationMot \neq up))$$
$$\wedge\ \mathbf{AG}\ (BottomPosition \Rightarrow \mathbf{AX}\ (TableElevationMot \neq down))$$
$$\wedge\ \mathbf{AG}\ (MaxRotation \Rightarrow \mathbf{AX}\ (TableRotationMot \neq clockwise))$$
$$\wedge\ \mathbf{AG}\ (MinRotation \Rightarrow \mathbf{AX}\ (TableRotationMot \neq counterClockwise))$$

Safety of the Robot   The safety property of the robot is split into several sub-requirements, so we acquire a conjunct of several CTL sub-formula.

1. *The robot never rotates over its bounds.*
   Similar to the requirements of the table, we have to determine what the critical bounds are and which kind of motor behaviour should be considered. The rotation bounds are determined by the angle. The rightmost position is *Arm1ToPress*; if this position is reached the rotation motor should not rotate counterclockwise anymore. The leftmost value is *Arm1ToTable*; in order not to cross this boundary the rotation motor should stop rotating clockwise.

$$\mathbf{AG}\ (angle = Arm1ToPress \Rightarrow \mathbf{AX}\ (RobotRotationMot \neq counterClockwise))$$
$$\wedge\ \mathbf{AG}\ (angle = Arm1ToTable \Rightarrow \mathbf{AX}\ (RobotRotationMot \neq clockwise))$$

2. *The robot never crashes with the press.*
   A crash with the press can be caused if either one of the robot arm motors or the rotation motor of the robot is running. In the former case, we have to ensure that, in the unloading phase, the press is in the right position (unloading position); similarly we have to ensure that, in the loading phase, the press has to be in load position.

   For unloading the press, **arm_2** is used and for loading the press, **arm_1**. That is, the motor *Arm2Mot* or *Arm1Mot* respectively, is *extending* and the angle of the robot has the right value. The unload position (load position) of the press is satisfied if the rotation motor of the robot and the motor of the press are both idle, the press is loaded (not loaded), and the press has reached the bottom position (middle position). The two cases are conjoined:

$$\mathbf{AG}\,(Arm1Mot = extend \,\wedge\, Angle = Arm1ToPress$$
$$\Rightarrow RobotRotationMot = idle \,\wedge\, PressMot = idle$$
$$\wedge\, \neg PressLoaded \,\wedge\, MiddlePositionPress)$$
$$\wedge\, \mathbf{AG}\,(Arm2Mot = extend \,\wedge\, Angle = Arm2ToPress$$
$$\Rightarrow RobotRotationMot = idle \,\wedge\, PressMot = idle$$
$$\wedge\, PressLoaded \,\wedge\, BottomPositionPress)$$

If the rotation motor works, we have to assure that the arms are retracted. Otherwise, the arms might be pushed into the press. The geometry of the production cell yields the following formula:

$$\mathbf{AG}\,(Angle = Arm2ToDepBelt \,\wedge\, RobotRotationMot = clockwise$$
$$\Rightarrow Arm2Ext = retracted)$$
$$\wedge\, \mathbf{AG}\,(Angle = Arm1ToTable \,\wedge\, RobotRotationMot = counterClockwise$$
$$\Rightarrow Arm2Ext = retracted)$$
$$\wedge\, \mathbf{AG}\,(Angle = Arm2ToDepBelt \,\wedge\, RobotRotationMot = counterClockwise$$
$$\Rightarrow Arm1Ext = retracted)$$

3. *The loaded first arm (*arm_1*) is never moved over the loaded table if the latter is in unload position.*
   That is, whenever the table is in its unload position, arm_1, responsible for loading, is not moving towards the table. The table is in unload position if it is in the top position and is maximally rotated. Additionally, we have to state that it is loaded. The fact that arm_1 is loaded is described in terms of the angle of the robot, the motion of arm_1's extending motor and the magnet of arm_1. arm_1 is moved to the table if the external function *Angle* has the value *Arm1ToTable*, and the motor is extending arm_1. Furthermore, the arm is loaded if the magnet is on. We get the following CTL-formula:

$$\mathbf{AG}\,(TopPosition \,\wedge\, MaxRotation \,\wedge\, TableLoaded$$
$$\Rightarrow \neg(Angle = Arm1ToTable \,\wedge\, Arm1Mot = extend$$
$$\wedge\, Arm1Magnet = on)$$

4. *The loaded second arm (*arm_2*) is never moved over the loaded deposit belt.*
   Similar to the last property, we use the same arguments: whenever the deposit belt is loaded, arm_2, responsible for loading the belt, should not be moved to it. The deposit belt is loaded if the state variable *DepositBeltReadyForLoading* is false. To describe the motion of arm_2, we have to consider the angle of the robot, the behaviour of arm_2's extending motor, and the magnet of arm_2. arm_2 is moved to the belt if the external function *Angle* has the value *Arm2ToDepBelt*, and the motor is extending arm_2. We model that the arm is loaded by the magnet of the arm being on. We get the following formula:

$$\mathbf{AG}\,(\neg DepositBeltReadyForLoading$$
$$\Rightarrow \neg(Angle = Arm2ToDepBelt \,\wedge\, Arm2Mot = extend$$
$$Arm2Mag = on))$$

5. *The robot never drops pieces outside safe areas.*
   That is, pieces should only be dropped when the arm is in a *safe* area: arm_1 should put every piece that it grabbed into the press, arm_2 should deposit the pieces on the belt. Consequently, we distinguish two cases: *whenever* arm_1

drops a piece, it has be in the press and *whenever* **arm_2** *drops a piece, it has to be over the belt.*

Dropping a piece is modelled through the magnet of an arm: if the magnet is on, and in the next state the magnet is off then the piece is dropped. It is sufficient to express the existence of such a next state using the existential quantifier for paths **E**.

arm_1 is in the press if its extension is appropriate and the angle of rotation has the right value. Similarly, we can express that **arm_2** is over the deposit belt. The following formula models the requirements:

$$\mathbf{AG}\,(Arm1Mag = on \;\wedge\; (\mathbf{EX}\,Arm1Mag = \mathit{off})$$
$$\Rightarrow Arm1Ext = Arm1IntoPress \;\wedge\; Angle = Arm1ToPress)$$

and

$$\mathbf{AG}\,(Arm2Mag = on \;\wedge\; (\mathbf{EX}\,Arm2Mag = \mathit{off})$$
$$\Rightarrow Arm2Ext = OverDepBelt \;\wedge\; Angle = Arm2ToDepBelt)$$

6. *A new blank will be put on the deposit belt only if there is enough space on the belt itself.*
   arm_2 puts a new blank on the belt if it is in the right position (right extension and right angle of the robot) and there is a next state where the magnet of **arm_2** is switched off. Moreover, there is enough space on the belt if the belt is ready for loading.

$$\mathbf{AG}\,(Arm2Mag = on \;\wedge\; Angle = Arm2ToDepBelt$$
$$\wedge\; Arm2Ext = OverDepBelt \;\wedge\; (\mathbf{EX}\,Arm2Mag = \mathit{off})$$
$$\Rightarrow DepositBeltReadyForLoading)$$

7. *The robot doesn't put blanks on the press if it is already loaded.*
   The press is already loaded if the variable *PressLoaded* is true. Whenever **arm_1** is in the right position (i.e., *Angle* equals *Arm1IntoPress* and **arm_1** is extended) and its magnet is on, a blank is put into the press if in a next state (at least one possible next state) the magnet switches off. We get the following implication:

$$\mathbf{AG}\,(Arm1Mag = on \;\wedge\; Angle = Arm1ToPress$$
$$\wedge\; Arm1Ext = Arm1IntoPress \;\wedge\; (\mathbf{EX}\,Arm1Mag = \mathit{off})$$
$$\Rightarrow \neg PressLoaded)$$

Safety of the Press   The safety requirements for the press consist of the following two points:

1. *The press is not moved downward if it is in the bottom position; it is not moved upward if it is in the top position*
   This is easily described using the state variables that model the motion of the press motor and the oracle functions for the position of the press: if the bottom position is reached, the press motor should not move downward; similarly, if the top position is reached, the press motor should not move upward.

We have to bear in mind that the condition of reaching the top or bottom position is also a condition for switching the motor to idle. Therefore, we profit by the next step quantification of CTL stipulating that whenever the boundary is reached in *all possible next states*, the motor should not move the press in this direction[1]. We get:

$$\textbf{AG}\,(bottomPositionPress \Rightarrow \textbf{AX}\,(PressMot \neq down))$$
$$\wedge\;\textbf{AG}\,(topPositionPress \Rightarrow \textbf{AX}\,(PressMot \neq up))$$

2. *The press only closes when there is no robot arm inside it.*
   The press is closing when the motor is running *up*. In this case, neither **arm_1** nor **arm_2** should ever be in the press. In terms of the ASM press model, the arms are in the press if the angle has the value *Arm1ToPress* or *Arm2ToPress* and the arm is completely extended. This leads to the following formula:

$$\textbf{AG}\,(PressMot = up$$
$$\Rightarrow \neg(Arm1Ext = Arm1IntoPress \;\wedge\; Angle = Arm1ToPress)$$
$$\wedge\;\neg(Arm2Ext = Arm2IntoPress \;\wedge\; Angle = Arm2ToPress))$$

Safety of the Deposit Belt   For the safety of the deposit belt it suffices to show that *the deposit belt is stopped when a blank reaches the end section of the belt itself and is not re-started until the travelling crane has picked up the piece.*

Whenever a piece reaches the end of the deposit belt, the deposit belt motor should stop running. This must happen in every next state[2]. For this we get the formula:

$$\textbf{AG}\,(PieceAtDepBeltEnd \Rightarrow \textbf{AX}\,(DepBeltMot = idle))$$

Safety of the Travelling Crane   To show the safety of the travelling crane we have to state four safety properties.

1. *The gripper always remains between the two (un)loading positions over the feed belt (loading position) and over the deposit belt (unloading position).*
   To express this safe behaviour of the crane, we regard the horizontal movement of the mechanics. Whenever the boundary of horizontal position is reached the *CraneHorizontalMot* should stop or move the gripper in the other direction. The boundaries are modelled by *GripperOverFeedBelt* and *GripperOverDepBelt*. We require that, when at these boundaries, in every possible next state, the horizontal motor does not move *toFeedBelt* and *toDepBelt* respectively. We get

$$\textbf{AG}\,(GripperOverFeedBelt$$
$$\Rightarrow \textbf{AX}\,(CraneHorizontalMot \neq toFeedBelt))$$

and

$$\textbf{AG}\,(GripperOverDepBelt$$
$$\Rightarrow \textbf{AX}\,(CraneHorizontalMot \neq toDepBelt))$$

2. *The gripper never crashes with the belts or the fixed part of the travelling crane.*
   Now we consider the vertical motion of the mechanics. Whenever a vertical boundary is reached, the motor should stop or move in the other direction in all reachable next states. We get four cases:

---

[1] In the real-life system, we have to require that the boundary be set with a sufficient margin in order to be able to stop the motor at a safe distance to prevent a collision.
[2] Again, in the real-life system, we have to require that the boundary be set with a sufficient margin in order to be able to stop the motor at a safe distance to prevent a collision.

- If the gripper's vertical position is a safe distance from the feed belt, then the motor should not move up in order not to push the gripper against the crane.

$$\mathbf{AG}\ (GripperVerticalPos = SafeDistanceFromFeedBelt$$
$$\Rightarrow \mathbf{AX}\ (CraneVerticalMot \neq up))$$

- If the gripper's vertical position is *OnDepositBelt* which is the lowest vertical position that is allowed then the crane should not move down any more.

$$\mathbf{AG}\ (GripperVerticalPos = OnDepositBelt$$
$$\Rightarrow \mathbf{AX}\ (CraneVerticalMot \neq down))$$

- If the gripper is over the feed belt then its lowest vertical position should be *OnFeedBelt*.

$$\mathbf{AG}\ (GripperVerticalPos = OnFeedBelt\ \wedge\ GripperOverFeedBelt$$
$$\Rightarrow \mathbf{AX}\ (CraneVerticalMot \neq down))$$

- If the vertical position is at its lowest point (i.e. *OnDepositBelt*) the horizontal motor should be idle in order to avoid pushing the gripper against the belt[3].

$$\mathbf{AG}\ (GripperVerticalPos = OnDepositBelt$$
$$\Rightarrow CraneHorizontalMot = idle)$$

3. *The crane does not drop pieces outside the safe area over the feed belt.*
The crane magnet drops a piece if it is on in a state and off in a next state. So we claim, if the magnet is on (i.e., gripper holds a piece) and the gripper is not in the right position (i.e., not over the feed belt with the right vertical position) then the magnet should not be switched off in the next state.

$$\mathbf{AG}\ (CraneMagnet = on$$
$$\wedge\ \neg(GripperOverFeedBelt\ \wedge\ GripperVerticalPos = OnFeedBelt)$$
$$\Rightarrow \mathbf{AX}\ (CraneMagnet \neq off))$$

4. *A new blank is put on the feed belt only if there is enough space on the latter to deposit the blank safely.*
To put a piece on the feed belt requires that the gripper is in the right position (vertically and horizontally) and the magnet is on. Furthermore, the ASM transition rule requires in its guard that the crane's vertical motor is moving down. If this is the case then the feed belt has to be free (i.e., *FeedBeltFree* is true).

$$\mathbf{AG}\ (GripperOverFeedBelt\ \wedge\ GripperVerticalPos = OnFeedBelt$$
$$\wedge\ CraneVerticalMot = down\ \wedge\ CraneMagnet = on$$
$$\Rightarrow FeedBeltFree)$$

The behaviour of the external functions (modelling the sensor values) which determine the behaviour of the modules has to be specified to reflect a reasonable system environment.

The model checker concluded that the conjunction of all safety formulas is satisfied by the SMV model of the Production Cell. The output shows the resources

---

[3]Since this condition is not a guard for stopping the horizontal motor, we can require this for the same state instead of all next states.

used:

```
resources used:
user time:  3.8 s, system time:  1.03333 s
BDD nodes allocated:  43529
Bytes allocated:  18087936
BDD nodes representing transition relation:  25718 + 1
```

A.3.4   The Liveness Requirements

Treating a blank is described in an abstract manner: e.g., a blank is transported by the feed belt if the feed belt changes from *NormalRun* to *CriticalRun*. We formalise the progress by means of CTL formulas of the following form

$$\mathbf{AG}\,(action\_executable \Rightarrow \mathbf{AF}\,(next\_action\_executable))$$

Following the semantics of CTL, this is satisfied if it is always (in every path in every state) the case that once *action* is executable then eventually *next_action* will be executable. We consider all pairs of actions and their successor actions, that are relevant to get the closed sequence of steps for the whole process, and get a set of CTL-formulas of the form discussed above. The liveness property is considered to be the conjunction of all progress formulas.

We suggest the following action/next action pairs with their respective formulas:

- Whenever *FeedBelt* is in *NormalRun* then eventually *FeedBelt* will reach *CriticalRun* (in the concrete model *NormalRun* and *CriticalRun* are introduced as macros whose expansions we use below):

$$\mathbf{AG}\,((FeedBeltMot = on \;\wedge\; \neg Delivering)$$
$$\Rightarrow \mathbf{AF}\,(FeedBeltMot = on \;\wedge\; Delivering))$$

- whenever *FeedBelt* is in *CriticalRun* then eventually *Table* will be stopped in load position:

$$\mathbf{AG}\,((FeedBeltMot = on \;\wedge\; Delivering)$$
$$\Rightarrow \mathbf{AF}\,(BottomPosition \;\wedge\; MinRotation \;\wedge$$
$$TableElevationMot = idle \;\wedge\; TableRotationMot = idle))$$

- whenever *Table* is stopped in load position then eventually *Table* will be stopped in unload position:

$$\mathbf{AG}\,((BottomPosition \;\wedge\; MinRotation \;\wedge$$
$$TableElevationMot = idle \;\wedge\; TableRotationMot = idle)$$
$$\Rightarrow \mathbf{AF}\,(TopPosition \;\wedge\; MaxRotation \;\wedge$$
$$TableElevationMot = idle \;\wedge\; TableRotationMot = idle))$$

- whenever *Table* is stopped in unload position then eventually *Robot* will unload the table; unloading the table is executed when *Arm1Mag* is switched to *on*, therefore, we use the guard of the respective transition rule as a condition for execution:

$$\mathbf{AG}\,((TopPosition \;\wedge\; MaxRotation \;\wedge$$
$$TableElevationMot = idle \;\wedge\; TableRotationMot = idle)$$
$$\Rightarrow \mathbf{AF}\,(Angle = Arm1ToTable \;\wedge\; Arm1Mot = extend$$
$$\wedge\; Arm1Ext = OverTable))$$

- whenever *Robot* has unloaded the table then eventually *Robot* will load the press; as above we we use the guard of the transition rule that has the effect that $Arm1Mag$ is switched *off*:

$$\mathbf{AG}\,((Angle = Arm1ToTable \;\wedge\; Arm1Mot = extend \\ \wedge\; Arm1Ext = OverTable) \\ \Rightarrow \mathbf{AF}\,(Angle = Arm1ToPress \;\wedge\; Arm1Mot = extend \\ \wedge\; Arm1Ext = Arm1IntoPress))$$

- whenever *Robot* has loaded the press then eventually *Press* will be *OpenForLoading*; *OpenForLoading* is defined to be a macros, here we use the macro definition[4]:

$$\mathbf{AG}\,((Angle = Arm1ToPress \;\wedge\; Arm1Mot = extend \\ \wedge\; Arm1Ext = Arm1IntoPress) \\ \Rightarrow \mathbf{AF}\,(MiddlePositionPress \;\wedge\; PressMot = idle))$$

- whenever *Press* is *OpenForLoading* then eventually *Press* will be *ClosedForForging*; *ClosedForForging* is defined as a macro, we unfold the definition:

$$\mathbf{AG}\,((MiddlePositionPress \;\wedge\; PressMot = idle) \\ \Rightarrow \mathbf{AF}\,(TopPositionPress \;\wedge\; PressMot = idle))$$

- whenever *Press* is *ClosedForForging* then eventually *Press* will be *OpenForUnloading*:

$$\mathbf{AG}\,((TopPositionPress \;\wedge\; PressMot = idle) \\ \Rightarrow \mathbf{AF}\,(BottomPositionPress \;\wedge\; PressMot = idle))$$

- whenever *Press* is *OpenForUnloading* then eventually *Robot* will unload the press; the press is unloaded if $Arm2Mag$ is switched to *on*, we consider the guard of the respective transition rule;

$$\mathbf{AG}\,((BottomPositionPress \;\wedge\; PressMot = idle) \\ \Rightarrow \mathbf{AF}\,(Angle = Arm2ToPress \;\wedge\; Arm2Mot = extend \\ \wedge\; Arm2Ext = Arm2IntoPress))$$

- whenever *Robot* has unloaded the press then eventually *Robot* will unload the DepositBelt; the DepositBelt is unloaded if $Arm2Mag$ is switched to *off*, again we consider the guard of the respective transition rule;

$$\mathbf{AG}\,((Angle = Arm2ToPress \;\wedge\; Arm2Mot = extend \\ \wedge\; Arm2Ext = Arm2IntoPress) \\ \Rightarrow \mathbf{AF}\,(Angle = Arm2ToDepBelt \;\wedge\; Arm2Mot = extend \\ \wedge\; Arm2Ext = OverDepBelt))$$

- whenever *Robot* has unloaded the DepositBelt then eventually *DepositBelt* will be in *NormalRun* (which is defined to be a macro):

$$\mathbf{AG}\,((Angle = Arm2ToDepBelt \;\wedge\; Arm2Mot = extend \\ \wedge\; Arm2Ext = OverDepBelt) \\ \Rightarrow \mathbf{AF}\,(DepBeltMot = run \;\wedge\; \neg Critical))$$

---

[4]In the following, we change the identifiers for the press position in order to distinguish them from the variables for the position of the table.

- whenever $DepositBelt$ is in $NormalRun$ then eventually $DepositBelt$ will be in $CriticalRun$ (which is defined to be a macro as well):

$$\mathbf{AG}\,((DepBeltMot = run \;\wedge\; \neg Critical)$$
$$\Rightarrow \mathbf{AF}\,(DepBeltMot = run \;\wedge\; Critical))$$

- whenever $DepositBelt$ is in $CriticalRun$ then eventually $Crane$ will unload the DepositBelt; the DepositBelt is unloaded if $CraneMagnet$ is switched to *on*, we regard the respective transition condition:

$$\mathbf{AG}\,((DepBeltMot = run \;\wedge\; Critical)$$
$$\Rightarrow \mathbf{AF}\,(GripperOverDepositBelt$$
$$\wedge\; GripperVerticalPos = OnDepositBelt$$
$$\wedge\; CraneHorizontalMot = idle \;\wedge\; CraneVerticalMot = idle$$
$$\wedge\; CraneMagnet = off\,))$$

- whenever $Crane$ has unloaded the DepositBelt then eventually the $Crane$ will load the FeedBelt; the FeedBelt is loaded when the $CraneMagnet$ is switched to *off*, we regard the respective transition condition:

$$\mathbf{AG}\,((GripperOverDepositBelt \;\wedge\; GripperVerticalPos = OnDepositBelt$$
$$\wedge\; CraneHorizontalMot = idle \;\wedge\; CraneVerticalMot = idle$$
$$\wedge\; CraneMagnet = off\,)$$
$$\Rightarrow \mathbf{AF}\,(CraneVerticalMot = down$$
$$\wedge\; GripperVerticalPos = OnFeedBelt \;\wedge\; GripperOverFeedBelt))$$

- whenever the $Crane$ has loaded the FeedBelt then eventually $FeedBelt$ will be in $NormalRun$;

$$\mathbf{AG}\,((CraneVerticalMot = down$$
$$\wedge\; GripperVerticalPos = OnFeedBelt \;\wedge\; GripperOverFeedBelt)$$
$$\Rightarrow \mathbf{AF}\,(FeedBeltMot = on \;\wedge\; \neg Delivering))$$

Results for the Liveness Property   We checked the requirement specification, as it is shown above, against the concrete model of the Production Cell. As a result we found that the system satisfies the property if (and only if) the initial condition is changed[5]. This is due to the fact, that the system is optimised in such a way that the robot can act only if at least two blanks are in the system: one to take out of the press and an another to be loaded into the press. We change the following initial values:

variable	old initial value	new initial value
$PressLoaded$	$false$	$true$
$Arm1Mag$	off	on
$Angle$	$Arm1ToTable$	$Arm2ToPress$

Also we take as an additional condition that initially the variable $PieceInDepBeltLightBarrier$ is $false$.

Moreover, the system environment and its behaviour must be modelled. Otherwise, the result will be a failure where the given counterexample is an irrelevant case.

---

[5]As Mearelli ([Mea96]) remarked, the initial condition chosen in his thesis is not suitable to start a simulation of a run. Therefore, the liveness property will fail in the initial state!

With respect of the changes listed above, we get the output

```
resources used:
user time:  10.1333 s, system time:  0.95 s
BDD nodes allocated:  56832
Bytes allocated:  18350080
BDD nodes representing transition relation:  25718 + 1
```

We conclude that the system model with respect to the modelled assumptions satisfies the liveness property as it was claimed in the setting of task.

Bibliography

[BAMP81] M. Ben-Ari, Z. Manna, and A. Pnueli. The temporal logic of branching time. In *Proc. of 8th Annual Symposium on Principles of Programming Languages*, 1981.

[BCL91]   J. Burch, E. Clarke, and D. Long. Symbolic model checking with partitioned transition relations. In *Int. Conf. on Very Large Scale Integration*, 1991.

[BCL+94] J. Burch, E. Clarke, D. Long, K. McMillan, and D. Dill. Symbolic model checking for sequential circuit verification. *IEEE Transactions on Computer Aided Design of Integrated Circuits and Systems*, 13(4):401–424, April 1994.

[BCM+92] J.R. Burch, E.M. Clarke, K.L. McMillan, D.L. Dill, and L.J. Hwang. Symbolic model checking 10^{20} states and beyond. *Information and Computation*, 98(2):142–170, June 1992.

[BCMD90] J. Burch, E. Clarke, K. McMillan, and D. Dill. Sequential circuits verification using symbolic model checking. In *Proc. of 27th ASM/IEEE Design Autiomation Conference*, 1990.

[BD94]   J. Burch and D. Dill. Automatic verification of pipelined microprocessor control. In D.L. Dill, editor, *Proc. of Int. Conf. on Computer Aided Verification, CAV'94*, volume 818 of *LNCS*, pages 68–80. Springer-Verlag, 1994.

[Ber99]   S. Berezin. *Combining Model Checking and Theorem Proving in Harware Verification*. PhD thesis, Carnegie Mellon University, 1999.

[BGL+00] S. Bensalem, V. Ganesh, Y. Lakhneck, C. Munoz, S. Owre, H. Rueß, J. Rushby, V. Rusu, H. Saïdi, N. Shankar, E. Singerman, and A. Tiwari. An Overview of SAL. In *Proc. of the Fifth Langley Formal Methods Workshop*, 2000.

[BGR94]   E. Börger, Y. Gurevich, and D. Rosenzweig. The Bakery Algorithm: Yet Another Specification and Verification. In E. Börger, editor, *Specification and Validation Methods*. Oxford University Press, 1994.

[BH99]   R. Bharadwaj and C. Heitmeyer. Model checking complete requirements specifications using abstraction. *Automated Software Engineering*, 6(1):37–68, 1999.

[BM97]   E. Börger and L. Mearelli. Integrating ASMs into the software development life cycle. *J.UCS Journal for Universal Computer Science (special issue)*, 3(5):689–702, 1997.

[Bör95]   E. Börger. Why Use Evolving Algebras for Hardware and Software Engineering? In M. Bartosek, J. Staudek, and J. Wiederman, editors, *Proceedings of SOFSEM'95, 22nd Seminar on Current Trends in Theory and Practice of Informatics*, volume 1012 of *LNCS*, pages 236–271. Springer, 1995.

[Bör99]   E. Börger. High Level System Design and Analysis using Abstract State Machines. In D. Hutter and W. Stephan and P. Traverso and M. Ullmann, editor, *Current Trends in Applied Formal Methods (FM-Trends 98)*, volume 1641 of *LNCS*, pages 1–43. Springer-Verlag, 1999.

[Bry86]   R. E. Bryant. Graph-based algorithms for boolean function manipulation. *IEEE Transactions On Computers*, C-35(8), August 1986.

[BT98]   S. Balakrishnan and S. Tahar. Modeling and Formal Verification of a Commercial Microcontroller for Embedded System Applications. In *Proc. IEEE 10th International Conference on Microelectronics (ICM'98)*, pages 107–110, 1998.

[Cas99]   G. Del Castillo. Towards comprehensive tool support for Abstract State Machines: The ASM Workbench tool environment and architecture. In D. Hutter et al., editors, *Current Trends in Applied Formal Methods (FM-Trends 98)*, volume 1641 of *LNCS*, pages 311–325, 1999.

[Cas00]   G. Del Castillo. *The ASM Workbench.* PhD thesis, Department of Mathematics and Computer Science of Paderborn University, 2000.

[CCGR99]   A. Cimatti, E. Clarke, F. Giunchiglia, and M. Roveri. NuSMV: A new symbolic model verifier. In *Proc. of Int. Conf. on Computer Aided Verfication, CAV'99*, volume 1633 of *LNCS*, pages 495–499. Springer-Verlag, 1999.

[CCL$^+$97] E. Cerny, F. Corella, M. Langevin, X. Song, S. Tahar, and Z. Zhou. Automated verification with abstract state machines using Multiway Decision Graphs. In T. Kropf, editor, *Formal Hardware Verification: Methods and Systems in Comparison*, volume 1287 of *LNCS*, pages 79–113. Springer Verlag, 1997.

[CD88]   E. Clarke and I. A. Draghicescu. Expressibility results for linear-time and branching-time logics. In J. W. de Bakker, editor, *Linear Time, Branching Time, and Partial Order in Logics and Models for Concurrency*, volume 354 of *LNCS*, pages 428–437. Springer-Verlag, 1988.

[CGH94]   E. Clarke, O. Grumberg, and K. Hamaguchi. Another Look at LTL Model Checking. In D.L. Dill, editor, *Proc. of 6th Int. Conf. on Computer Aided Verification, CAV'94*, volume 818 of *LNCS*, pages 415–427. Springer-Verlag, 1994.

[CGH97]   E. Clarke, O. Grumberg, and H. Hamaguchi. Another Look at LTL Model Checking. *Formal Methods in System Design*, 10(1), February 1997.

[CGL92]   E. Clarke, O. Grumberg, and D. Long. Model-checking and abstraction. In *Proc. of the 19th ACM Symposium on Principles of Programming Languages*, pages 343–354. ACM Press, 1992.

[CGL94]   E. Clarke, O. Grumberg, and D. Long. Verification tools for finite-state concurrent systems. In *A Decade of Concurrency - Reflections and Perspectives*, volume 803 of *LNCS*. Springer-Verlag, 1994.

[CGP00]   E. Clarke, O. Grumberg, and D. Peled. *Model Checking*. MIT Press, 2000.

[CH85]   A. B. Cremers and T. N. Hibbard. *VLSI and Modern Signal Processing*, chapter Executable Specification of Concurrent Algorithms in Terms of Applicative Data Space Notation, pages 201–223. Prentice-Hall, 1985.

[CMP92]   E. Chang, Z Manna, and A. Pnueli. The safety-progress classification. Technical Report STAN-CS-92-1408, Department of Computer Science, Stanford University, 1992.

[CN94]   D Cyrluk and P. Narendran. Ground temporal logic: A logic for hardware verification. In D. Dill, editor, *Proc. of Int. Conf. on Computer Aided Verification, CAV'94*, volume 818 of *LNCS*, pages 247–259. Springer-Verlag, 1994.

[CW00]   G. Del Castillo and K. Winter. Model checking support for the ASM high-level language. In S. Graf and M. Schwartzbach, editors, *Proc. of 6th Int. Conference for Tools and Algorithms for the Construction and Analysis of Systems, TACAS 2000*, volume 1785 of *LNCS*, pages 331–346. Springer-Verlag, 2000.

[CZS$^+$97] F. Corella, Z. Zhou, X. Song, M. Langevin, and E. Cerny. Multiway Decision Graphs for automated hardware verification. *Formal Methods in System Design*, 10(1), 1997.

[Day93]   N. Day. A model checker for statecharts (linking case tools with formal methods). Technical Report 93 - 35, Dep. of Computer Science, Univ. of British Columbia, Vancouver, B.C., Canada, October 1993.

[DB98a]   R. Drechsler and B. Becker. *Binary Decision Diagrams*. Kluwer Academic Publishers, 1998.

[DB98b]   R. Drechsler and B. Becker. *Graphenbasierte Funktionsdarstellung*. Teubner, 1998.

[Dur98]   A. Durand. Modeling cache coherence protocol - a case study with FLASH. In U. Glässer and P. Schmitt, editors, *Proc. of the 28th Annual Conference of the German Society of Computer Science*, Technical Report, Magdeburg University, Germany, 1998.

[EH86]   E. Allen Emerson and Joseph Y. Halpern. "Sometimes" and "Not Never" revisited: On branching versus linear time temporal logic. *Journal of the ACM*, 33(1):151–178, January 1986.

[Eme90]   E. Allen Emerson. Temporal and modal logic. In J. van Leeuwen, editor, *Handbook of Theoretical Coomputer Science*, volume B, pages 996–1072. Elsevier Science Publishers, Amsterdam, 1990.

[F$^+$96]   T. Filkorn et al. *SVE Users' Guide*. Siemens AG, München, 1996.

[FGM$^+$92] J.-C. Fernandez, H. Garavel, L. Mounier, A. Rasse, C. Rodrguez, and J. Sifakis. A toolbox for the verification of LOTOS programs. In *Proc. of the 14th Int. Conf. on Software Engineering, (ICSE'14)*, pages 246–259. ACM, 1992.

[For96]   Formal Systems (Europe) Ltd. *Failure Divergence Refinement, FDR 2.0, User Manual*, August 1996.

[GB96]   B. Grahlmann and E. Best. Pep – more than a petri net tool. In
            T. Magaria and B. Steffen, editors, *Proc. of the 2th Int. Conference for
            Tools and Algorithms for the Construction and Analysis of Systems,
            TACAS'96*, volume 1055 of *LNCS*, pages 397–401. Springer-Verlag,
            1996.

[GPVW95]   Rob Gerth, Doron Peled, Moshe Y. Vardi, and Pierre Wolper. Simple
            On-the-fly Automatic Verification of Linear Temporal Logic. In *Proc. of
            the 15th Work. Protocol Specification, Testing, and Verification*, War-
            saw, June 1995. North-Holland.

[GR97]   M. Große-Rhode. Transition specifications for dynamic abstract data
            types. *Applied Categorical Structures*, 5:265–308, 1997.

[GR01]   A. Gargantini and E. Ricobene. ASM-based testing: Coverage crite-
            ria and automatic tests generation. In *Proc. of ASM Workshop 2001*,
            LNCS. Springer-Verlag, 2001. To appear.

[Gro96]   The VIS Group. VIS: A System for Verification and Synthesis. In
            R. Alur and T. Henzinger, editors, *Proc. of the 8th Int. Conf. on Com-
            puter Aided Verifaction, CAV'96*, volume 1102 of *LNCS*, pages 428–432.
            Springer-Verlag, 1996.

[GS97]   S. Graf and H. Saïdi. Construction of abstract state graphs with PVS.
            In O. Grumberg, editor, *Proc. of Int. Conf. on Computer Aided Veri-
            fication, CAV'97*, volume 1254 of *LNCS*, pages 72–83. Springer-Verlag,
            1997.

[Gur95]   Y. Gurevich. Evolving Algebras 1993: Lipari Guide. In E. Börger,
            editor, *Specification and Validation Methods*. Oxford University Press,
            1995.

[Gur97]   Y. Gurevich. May 1997 Draft of the ASM Guide. Technical report,
            University of Michigan EECS Department, 1997.

[HDB97]   R. Hojati, D. Dill, and R. Brayton. Verifying linear temporal properties
            of data insensitive controllers using finite instantiations. In *Proc. of Int.
            Conf. on Hardware Description Languages, CHDL'97*, IFIP, pages 60–
            73. Chapman & Hall, 1997.

[Hei95]   M. Heinrich. The FLASH protocol - version: 1.4. Technical report,
            Stanford University FLASH Group, 1995.

[HGD95]   H. Hungar, O. Grumberg, and W. Damm. What if model checking must
            be truly symbolic? In E. Brinksma, editor, *Workshop on Tools and
            Algorithms for the Construction and Analysis of Systems, TACAS'95*,
            volume 1019 of *LNCS*. Springer-Verlag, 1995.

[HHK96]   R.H. Hardin, Z. Har'El, and R.P. Kurshan. COSPAN. In T. Henzinger
            R. Alur, editor, *Proc. of the 8th Int. Conf. on Computer Aided Verifac-
            tion, CAV'96*, volume 1102 of *LNCS*, pages 423–427. Springer-Verlag,
            1996.

[Hol97]   Gerard J. Holzmann. The SPIN Model Checker. *IEEE Transactions
            on Software Engineering*, 23(5):279–295, May 1997.

[HS95]   K. Havelund and N. Shankar. Experiments in theorem proving and
            model checking for protocol verification. Technical report, SRI Inter-
            national Menlo Park, USA, 1995.

BIBLIOGRAPHY

[Hug]   J.K. Huggins. Abstract state machines home page. EECS Department,
            University of Michigan. http://www.eecs.umich.edu/gasm/.

[KOH⁺94]   J. Kuskin, D. Ofelt, M. Heinrich, J. Heinlein, R. Simoni, K. Ghara-
            chorloo, J. Chapin, D. Nakahira, J. Baxter, M. Horowitz, A. Gupta,
            M. Rosenblum, and J. Hennessy. The stanford FLASH multiprocessor.
            In *21th Int. Symp. on Computer Architecture*, Computer Archtiecture
            News, pages 302–312, 1994.

[Kro98]   T. Kropf. Hardware verifikation. Vorlesungsskript, WS 98/99, 1998.

[KW97]   E. Kindler and R. Walter. Mutex needs fairness. *Information Processing
            Letters*, 62, 1997.

[Lam80]   Leslie Lamport. "Sometimes" is sometimes "Not Never". In *Proc. of the
            7th Annual ACM Symposium on Principles of Programming Languages*,
            pages 174–185. ACM, 1980.

[LEW96]   J. Loeckx, H.-D. Ehrich, and M. Wolf. *Specifiaction of Abstract Data
            Types*. Wiley, Teubner, 1996.

[LGS⁺95]   C. Loiseaux, S. Graf, J. Sifakis, A. Bouajjani, and S. Bensalem. Prop-
            erty preserving abstractions for the verification of concurrent systems.
            Formal Methods in System Design, 6(1), 1995.

[LL95]   C. Lewerentz and T. Lindner. *Formal Developement Of Reactive Sys-
            tems. Case Study "Production Cell"*, volume 891 of *LNCS*. Springer-
            Verlag, 1995.

[Lon93]   David E. Long. *Model Checking, Abstraction and Compositional Veri-
            fication*. PhD thesis, Carnegie Mellon University, 1993.

[LP85]   O. Lichtenstein and A. Pnueli. Checking that finite state concurrent
            programs satisfy their linear specification. In *Proc. of the 12th Annual
            ACM Symposium on Principles of Programming Languages*. ACM, New
            York, January 1985.

[LR98]   P. Liggesmeyer and M. Rothfelder. Towards automated proof of fail-safe
            behavior. In W. Ehrenberger, editor, *Proc. of Int. Conf. on Computer
            Safety, Reliablity and Security, SAFECOMP'98*, volume 1516 of *LNCS*,
            pages 169–184. Springer-Verlag, 1998.

[LTZ⁺96]   M. Langevin, S. Tahar, Z. Zhou, X. Song, and E. Cerny. Behavioral
            verification of an ATM switch fabric using implicit abstract state enu-
            meration. In *Proc. of IEEE Int. Conf. on Computer Design (ICCD'96)*,
            pages 20–26. IEEE Computer Society Press, 1996.

[McM93]   K. McMillan. *Symbolic Model Checking*. Kluwer Academic Publishers,
            1993.

[Mea96]   L. Mearelli. An Evolving Algebra Model Of The Production Cell. Mas-
            ter's thesis, Universita di Pisa, 1996.

[Mil]   K. Mc Millan. Ken McMillan's home page at UCB. http://www-
            cad.eecs.berkeley.edu/~kenmcmil.

[MMP98]   A. Merceron, M. Müllerburg, and G.M. Pinna. Verifying a time-triggered protocol in a multi-language environment. In W. Ehrenberger, editor, *Proc. of Int. Conf. on Computer Safety, Reliablity and Security, SAFECOMP'98*, volume 1516 of *LNCS*, pages 185–195. Springer-Verlag, 1998.

[Mor94]   C. Morgan. *Programming from Specifications*. Prentice Hall, 1994.

[MS]   Faron Moller and Perdita Stevens. Edinburgh Concurrency Workbench user manual (version 7.1). Available from http://www.dcs.ed.ac.uk/home/cwb/.

[MSC97]   O. Ait Mohamed, X. Song, and E. Cerny. On the Non-Termination of MDG-based abstract state enumeration. In *Proc. of the IFIP Advanced Research Working Conference on Correct Hardware Design and Verification Methods (Charme'97)*, pages 218–235, 1997.

[ORS92]   S. Owre, J.M. Rushby, and N. Shankar. PVS: A prototype verification system. In D. Kapur, editor, *Proc. of Int. Conf. on Automated Deduction, CADE 11*, volume 607 of *LNAI*, pages 748–752. Springer-Verlag, 1992.

[Plo00]   C. Plonka. Model checking for the design with abstract state machines. Master's thesis, University of Ulm, Department of Applied Information Processing, 2000.

[Ros94]   A. W. Roscoe. Model-Checking CSP. In *A classical Mind, Essays in Honour of C.A.R. Hoare*. Prentice Hall, 1994.

[Sha00]   N. Shankar. Symbolic Analysis of Transition Systems. In Y. Gurevich, P. Kutter, M. Odersky, and L. Thiele, editors, *Abstract State Machines, Theory and Applications, ASM 2000*, volume 1912 of *LNCS*, pages 287–302. Springer-Verlag, 2000.

[SMT01]   M. Shirazipour, Y. Mokhtari, and S. Tahar. Model checking and refinement of ASM models using SMV. Technical report, Concordia University, Faculty of Engineering and Computer Science, 2001.

[Spi99]   M. Spielmann. Automatic verification of abstract state machines. In N. Halbwachs and D. Peled, editors, *Proc. of Int. Conf. on Computer Aided Verification, CAV '99*, volume 1633 of *LNCS*, pages 431–442. Springer-Verlag, 1999.

[SS99]   H. Saïdi and N. Shankar. Abstract and model check while you prove. In N. Halbwachs and D. Peled, editors, *Proc. of Int. Conf. on Computer Aided Verification, CAV'99*, volume 1633 of *LNCS*, pages 443–453. Springer-Verlag, 1999.

[vL90]   Jan van Leeuwen, editor. *Formal Models and Semantics*, volume B. Elsevier, 1990.

[VW86]   M. Y. Vardi and P. Wolper. An automata-theoretic approach to automatic program verification. In *Proc. of the First Annual Symposium on Logic in Computer Science*, pages 322–331. IEEE Computer Society Press, 1986.

[Win97]   K. Winter. Model Checking for Abstract State Machines. *J.UCS Journal for Universal Computer Science (special issue)*, 3(5):689–702, 1997.

BIBLIOGRAPHY

[Win00]   K. Winter. Towards a Methodology for Model Checking ASM: Lessons
           learned from the FLASH Case Study. In Y. Gurevich, P. Kutter,
           M. Odersky, and L. Thiele, editors, *Abstract State Machines - Theory
           and Applications, Int. Workshop ASM'2000, Selected papers*, volume
           1912 of *LNCS*, pages 341–360. Springer-Verlag, 2000.

[Win01]   K. Winter. Model Checking ASM: the MDG approach. In *Proc. of
           ASM Workshop 2001*, LNCS. Springer-Verlag, 2001. To appear.

[Wol81]   P. Wolper. Temporal logic can be more expressive. In *Proc. of the
           22nd Annual Symposium on Foundations of Computer Sciences*, pages
           340–348. IEEE, 1981.

[Wol97]   Pierre Wolper. The meaning of formal: from weak to strong formal
           methods. *Springer International Journal on Software Tools for Tech-
           nology Transfer*, 1(1-2):6–8, 1997.

[WVF97]   Jeannette M. Wing and Mandana Vaziri-Farahani. A case study in
           model checking software systems. *Science of Computer Programming*,
           28:273–299, 1997.

[XCS+98]   Y. Xu, E. Cerny, X. Song, F. Corella, and O. Ait Mohamed. Model
           checking for a first-order temporal logic using Multiway Decision
           Graphs. In *Proc. of Int. Conf. on Computer Aided Verification
           (CAV'98)*, volume 1427 of *LNCS*, pages 219–231. Springer-Verlag, 1998.

[Xu99]   Y. Xu. *Model Checking for a First-order Temporal Logic Using Multi-
           way Decision Graphs*. PhD thesis, University of Montreal, 1999.

[Zho96]   Z. Zhou. *MDG Tools (V1.0) Developer's Manual*. IRO University of
           Montreal, Montreal, Canada, June 1996.

[ZSC+95]   Z. Zhou, X. Song, F. Corella, E. Cerny, and M. Langevin. Partitioning
           transition relations automatically and efficiently. In *IEEE Proceedings
           of Fifth Great Lakes Symposium on VLSI (GLSVLSI'95)*, March 1995.

[ZST+96]   Z. Zhou, X. Song, S. Tahar, F. Corella E. Cerny, and M. Langevin.
           Verification of the island tunnel controller using Multiway Decision
           Graphs. In M. Srivas and A. Camilleri, editors, *Proc. of Int. Conf.
           on Formal Methods in Computer-Aided Design (FMCAD'96)*, volume
           1166 of *LNCS*, pages 233–246. Springer-Verlag, 1996.

www.ingramcontent.com/pod-product-compliance
Lightning Source LLC
LaVergne TN
LVHW022312060326
832902LV00020B/3422